SPACE-TIME LAYERED INFORMATION PROCESSING FOR WIRELESS COMMUNICATIONS

SPACE-TIME LAYERED INFORMATION PROCESSING FOR WIRELESS COMMUNICATIONS

Mathini Sellathurai

Simon Haykin

Celebrating 125 Years
of Engineering the Future

WILEY

A JOHN WILEY & SONS, INC., PUBLICATION

For general information on our other products and services or for technical support, please contact our
Customer Care Department within the United States at (800) 762-2974, outside the United States at
(317) 572-3993 or fax (317) 572-4002.

Wiley also publishes its books in a variety of electronic formats. Some content that appears in print
may not be available in electronic formats. For more information about Wiley products, visit our web
site at www.wiley.com.

Library of Congress Cataloging-in-Publication Data:

Sellathurai, Mathini, 1968-
  Space-time layered information processing for wireless communications / by
Mathini Sellathurai and Simon Haykin.
       p. cm.
  Includes bibliographical references and index.
  ISBN 978-0-471-20921-8
  1. Space time codes. 2. MIMO systems. I. Haykin, Simon S., 1931- II. Title.
  TK5103.4877.S45 2009
  621.3840285′572–dc22

                                                        2009005657

Printed in the United States of America

10 9 8 7 6 5 4 3 2 1

CONTENTS

LIST OF TABLES

LIST OF FIGURES

1

INTRODUCTION

1.1 BRIEF HISTORICAL NOTES

In the last decade of the twentieth century, two groundbreaking ideas were published, which, in their own individual ways, have shaped many facets of digital communications and signal processing in both theoretical and practical terms.

The first idea on turbo codes was presented at the 1993 IEEE International Conference on Communications (ICC) that was held in Geneva, Switzerland, in May of that year. At that conference, Berrou, Glavieux, and Thitimajshima presented a paper entitled "Near Shannon Limit Error-Correcting Coding and Decoding: Turbo Codes," and with it the ever-expanding field of turbo-information processing was born [17].

Then, three years later, Foschini published a paper entitled "Layered Space-Time Architecture for Wireless Communication in a Fading Environment When Using Multi-Element Antennas" in the *Bell Laboratories Technical Journal* [43]. With the publication of this second paper, the ever-expanding field of multiple-input multiple-output (MIMO) wireless communications was born.

Although entirely different in their theory and applications, turbo-information processing and MIMO wireless communications, share two common points:

- They were both ideas conceived as a result of "thinking outside of the box" and were initially received with a skepticism by experts in the field.

- Since their invention in the 1990s, they have both evolved at an unprecedented pace, reaching a state of maturity in just over a decade.

The particular form of MIMO wireless communications described in Foschini's paper was named the "Bell Labs Layered Space-Time (BLAST)" architecture. With the early formulations of the two ideas, turbo processing and BLAST architecture, it was logical that these two ideas be combined into what we now refer to as "Turbo-BLAST," on which research was initiated when the first author of this book joined the senior author as a Ph.D. student in 1998. Indeed, it was Sellathurai's thesis, entitled "Turbo-BLAST, A Novel Technique for Multi-Transmit and Multi-Receive Wireless Communications," and subsequent publications that led to the writing of this book. Simply put, Turbo-BLAST offers the advantage of building a layered space-time wireless communication system that is both spectrally and computationally efficient.

1.2 TURBO-INFORMATION PROCESSING

The turbo-coding scheme, originally formulated by Berrou, Glavieux, and Thitimajshima, is a codec, in which the encoder and decoder distinguish themselves from the traditional codecs in two fundamental ways:

1. The encoder consists of two parallel constituent encoders with an interleaver between them, as depicted in Figure 1.1. The purpose of the interleaver is to randomize the incoming stream of bits to ensure that the respective inputs of the two constituent encoders are as dissimilar as practically possible.
2. Correspondingly, the decoder consists of two constituent decoders separated by an interleaver and a de-interleaver, forming a closed-loop feedback system in the manner depicted in Figure 1.2. The interleaver and de-interleaver are positioned inside the decoder in such a way that the inputs applied to

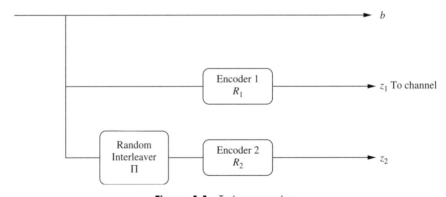

Figure 1.1 Turbo encoder.

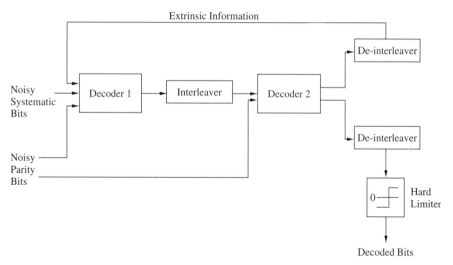

Figure 1.2 Turbo decoder.

each constituent decoder correspond to the pertinent constituent encoder. In particular, each constituent decoder operates on three different inputs:

- The systematically encoded (message) bits
- The parity-check bits associated with the systematic bits
- The information bits produced by the other constituent decoder about the likely values of the received message bits.

The turbo decoder of Figure 1.2 is an iterative decoder. An important novel feature of this decoder is the application of feedback around all of the components constituting the decoder. Another important feature that is equally novel, in its own way, is the notion of extrinsic information that is basic to the operation of the turbo decoder. The extrinsic information, generated by a decoding stage for a set of systematic (message) bits, is defined as the difference between the log-likelihood ratio computed at the output of that particular decoding stage and the intrinsic information represented by the log-likelihood ratio fed back to the input of the decoding stage. In effect, extrinsic information is the incremental information gained by exploiting the dependencies that exist between a specific message bit and the incoming raw data bits processed by the decoder. Thus, in a loose sense, we may view the role of extrinsic information in turbo decoding as the "error signal" in a conventional closed-loop feedback system.

The concept of turbo codes was originally conceived by Berrou, Glavieux, and Thitimajshima in the context of channel codes, with the primary purpose of approaching the Shannon limit in a computationally efficient manner. Today, this concept is being applied not only in channel coding, but also in source coding, joint source-channel coding, channel equalization, synchronization, and MIMO wireless communications. For an important survey of these applications to turbo-information processing, the reader is referred to the special issue of the *Proceedings of the IEEE*, vol. 95, July 2007 [141].

1.3 MIMO WIRELESS COMMUNICATIONS

In a wireless environment, the transmitted signal reaches the intended receiver via a multiplicity of propagation paths; hence, the resulting components of the wireless channel output may end up adding in a destructive manner. Such a situation may result in serious degradation in the performance of the wireless communication system. This multipath phenomenon is commonly referred to as channel fading. To overcome the degrading effects of channel fading, it is common practice to use diversity. The basic idea of this technique is to provide the receiver with a set of independently faded replicas of the transmitted signal in the hope that at least one of them will have been received in a reasonably correct manner.

Diversity can be realized in a variety of ways under one of three basic headings:

1. Diversity on receive
2. Diversity on transmit
3. Diversity on both transmit and receive

In MIMO wireless communications, it is the third form of diversity that is employed. Specifically, the transmitter employs an array of antenna elements, and the receiver employs another array of antenna elements of its own. These two antenna arrays may embody different numbers of antenna elements.

The interesting properties of a MIMO wireless communication system are summarized as follows:

1. Under certain environmental conditions, fading is viewed not as a nuisance, but rather as a possible environmental source of performance improvement.
2. The combined use of space diversity at both the transmit and receive ends of the MIMO wireless link may provide the basis for an increase in channel capacity or spectral efficiency of the system.
3. Unlike the use of conventional techniques to increase channel capacity, in MIMO wireless communications the increase in channel capacity is achieved by increasing computational complexity while, at the same time, keeping the primary communication resources (i.e., total transmit power and channel bandwidth) constant.

These are remarkable properties.

Figure 1.3 shows the block diagram of a MIMO wireless link, where n_t is the number of transmit antennas and n_r is the number of receive antennas. Suppose now we make two assumptions:

1. The wireless link is modeled as a narrowband flat-fading channel.

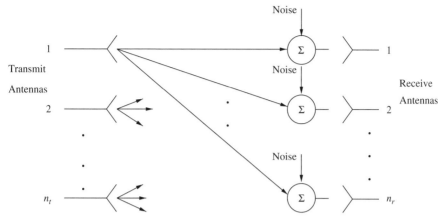

Figure 1.3 Schematic a of MIMO wireless link.

2. The number of transmit antennas and the number of receive antennas have a common value denoted by N.

Under these special conditions, we find that as N approaches infinity, the (ergodic) capacity of the MIMO channel grows asymptotically (at least) linearly with N, as shown by

$$\lim_{N \to \infty} \frac{C}{N} \geq \text{constant} \tag{1.1}$$

where C denotes the channel capacity. This asymptotic result teaches us that, by increasing the computational complexity of a MIMO wireless communication system through the use of multiple antennas at both the transmit and receive ends of a wireless link, we are able to increase the spectral efficiency of the link for more than is possible by conventional means (i.e., increasing the signal-to-noise ratio). Indeed, it is this important result that is responsible for the increasing interest in the deployment of MIMO wireless links.

1.4 ORGANIZATION OF THE BOOK

Two major motivations for MIMO wireless communication research exist. On the one hand, information theorists wish to understand the ultimate limits of bandwidth-efficient digital wireless communications system by exploiting the MIMO technology. They attempt to find techniques that attain Shannon's capacity limit. On the other hand, a communication engineer wishes to design techniques that are practically feasible and also to achieve a significant portion of the great capacity promised by information theory. The two motivations are certainly not mutually exclusive and are slowly converging to provide a more principled approach to MIMO wireless communication.

This book is concerned about both of these motivations. In particular, Chapter 2 presents the MIMO channel capacity limits and Chapter 3–6 describe unconstrained signaling techniques, exemplified by the BLAST architectures, whose aim is to increase the channel capacity by using standard channel codes.

Chapter 2 is devoted to the spectral efficiency of MIMO channels under various channel conditions. We provide the information theory concepts and capacity limits of MIMO channels over Rayleigh fast fading and quasi-static fading. In particular, we derive the MIMO channel capacity from the first principle assuming that the receiver has knowledge of the channel state. In this scenario, when the wireless communication environment is endowed with rich scattering, the information capacity of the wireless channel is roughly proportional to the number of transmit or receive antennas, whichever is smaller. That is to say, we have the potential to achieve a spectacular increase in spectral efficiency, with the channel capacity of the link being roughly doubled by doubling the number of antennas at both ends of the link.

In Chapter 3, we describe a family of MIMO wireless communication systems popularized as BLAST architectures. In particular, BLAST architectures use standard one-dimensional error-correction codes and low-complexity interference-cancellation schemes to construct and decode powerful two-dimensional space-time codes. These MIMO systems offer spectacular increases in spectral efficiency, provided that three conditions are met:

- The system operates in a rich scattering environment.
- Appropriate coding structures are used.
- Error-free decisions are available in the interference-cancellation schemes, which, in turn, assumes the combined use of arbitrarily long (and therefore powerful) error-correction codes and perfect decoding.

The material presented herein focuses on three specific implementations of BLAST, depending on the type of coding employed:

- Diagonally layered space-time architecture known as diagonal BLAST or simply D-BLAST, which provides the standard framework for MIMO wireless communications;
- A simplified version of BLAST known as vertical BLAST or V-BLAST, which is the first practical implementation of MIMO wireless communications demonstrating a spectral efficiency as high as 40 bits/s/Hz in real time with significant reduction in system complexity;
- Stratified D-BLAST.

In Chapter 4, we review the framework of the turbo principle and its applications in space-time channels. In particular, we describe serial and parallel concatenated turbo codes and their iterative decoders, soft-in/soft-out modules, which are exemplified by the BCJR algorithm that performs maximum *a posteriori* estimation on a bit-by-bit basis in the decoding of turbo codes and their lower complexity and numerically less sensitive approximations, the extraction of extrinsic information. The turbo decoding principle features prominantly in Chapters 5 and 6.

In Chapter 5, we describe the Turbo-BLAST architecture which is suitable for for high-throughput wireless communications, exploiting the following basic ideas:

- A random layered space-time coding scheme, which is based on the use of independent block coding and space-time interleaving.
- A turbo-like receiver, also known as an iterative interference cancellation and decoding receiver.

In particular, we show how the turbo principles applied to BLAST architectures can significantly improve overall system performance.

In Chapter 6, we discuss another important class of MIMO architecture, known as turbo-MIMO systems, where the layer code construction is based on space-time bit-interleaved coded modulation (ST-BICM). Turbo-MIMO is a highly effective system when used in conjunction with receivers employing iterative detection and decoding. This chapter also presents three recent low-complexity detection schemes:

- Minimum mean-squared based soft-interference cancellation;
- Low complexity implementations of sphere detections;
- Iterative tree search detection;
- List-sphere detection.

The chapter includes the analysis of flat and frequency-selective fading channels for Turbo-MIMO.

2

MIMO CHANNEL CAPACITY

2.1 INTRODUCTION

The major concern in wireless communications research is to provide techniques that use the frequency spectrum, which is a scarce resource, efficiently. The basic information theory result reported in a pioneering paper [45] by Foschini and Gans showed that great spectral efficiency can be achieved through the use of multiple-input, multiple-output (MIMO) wireless systems. The major conclusion of their work is that the capacity of a multiple-transmit, multiple-receive system far exceeds that of a single-antenna system. In particular, in a Rayleigh flat-fading environment, a MIMO link has an asymptotic capacity that increases linearly with the number of transmit and receive antennas, provided that the complex valued propagation coefficients between all pairs of transmit and receive antennas are statistically independent and known to the receiver.

This chapter presents a detailed discussion of the channel capacity limits of different MIMO wireless channels systems. In particular, we derive the MIMO channel capacity from first principles, assuming that the receiver has knowledge of the channel state. In this scenario, when the wireless communication environment is endowed with rich scattering, the information capacity of the wireless channel is roughly proportional to the number of transmit or receive antennas, whichever is less. That is to say, we have a spectacular increase in spectral efficiency, with the channel capacity of the link being roughly doubled by doubling the number of antennas at both ends of the link.

Space-Time Layered Information Processing for Wireless Communications,
By Mathini Sellathurai and Simon Haykin

Section 2.2 describes the notion of MIMO wireless communications and defines the notations for the baseband MIMO structure used in this book. Throughout the presentation, we adopt a baseband view of the transmitted and received signals, hence the use of complex signals and channels. Section 2.3 introduces some preliminaries of the notion of channel capacity and complex multidimensional Gaussian distribution. This is followed by Section 2.4 on the channel capacity of MIMO systems, assuming that the receiver has knowledge of the channel state. This includes the derivation of ergodic and outage capacity equations for MIMO wireless communications. Section 2.5 presents another viewpoint of the input-output relation of the MIMO channel by applying a transformation known as singular-value decomposition to the channel matrix; the resulting decomposition is insightful in the context of fading correlation. A summary and discussion are provided in Section 2.6.

2.2 MULTIPLE-INPUT, MULTIPLE-OUTPUT ANTENNA SYSTEMS

In this section, we discuss MIMO wireless communications, also referred to in the literature as multiple-transmit, multiple-receive (MTMR) wireless communications. MIMO wireless communications include space diversity on both transmit and receive ends. Figure 2.1 shows the block diagram of a MIMO wireless communication link. The signals transmitted by the n_t transmit antennas over the wireless channel are all chosen to lie inside a common frequency band. Naturally, the transmitted signals are scattered differently by the channel and received by all n_r receive antennas operating in the same frequency band.

In particular, MIMO has become a pervasive information theory concept that promises high spectral efficiencies for wireless communications in a fading environment. In theory, the spectral efficiency of a communication system is intimately linked to the channel capacity of the system. To evaluate the channel capacity of MIMO wireless communications, we begin by formulating a baseband channel model for the system, as described next.

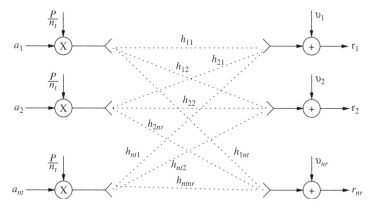

Figure 2.1   An (n_t, n_r) system.

2.2.1 Basic Baseband Channel Model

Consider a MIMO narrowband wireless communication system built around a flat-fading channel, with n_t transmit antennas and n_r receive antennas (Figure 2.1). The antenna configuration is hereafter referred to as the pair (n_t, n_r). For a statistical analysis of the MIMO system in what follows, we use baseband representations of the transmitted and received signals as well as the channel. In particular, we introduce the following notations:

- The spatial parameter

$$N = \min\{n_r, n_t\} \tag{2.1}$$

- The n_t-by-1 vector

$$\mathbf{x}(t) = [x_1(t), x_2(t), \ldots, x_{n_t}(t)]^T \tag{2.2}$$

  denotes the complex signal vector transmitted by the n_t antennas at discrete time t. The symbols constituting the vector $\mathbf{x}(t)$ are assumed to have zero mean and common variance σ_x^2 drawn from the modulation constellation set

$$\mathcal{X} = \{\tilde{x}_1, \tilde{x}_2, \ldots, \tilde{x}_{N_l}\}$$

  The total power of $\mathbf{x}$ is constrained to P regardless of n_t,

$$P = n_t \sigma_x^2 \tag{2.3}$$

  For P to be maintained constant, the variance (i.e., power radiated by each transmit antenna) must be inversely proportional to n_t.

- For the flat-fading and therefore memoryless channel, we may use h_{ik} to denote the sampled complex gain of the channel from transmit antenna k to receive antenna i at discrete time t, where $k = 1, 2, \ldots, n_t$ and $i = 1, 2, \ldots, n_r$. We may thus express the n_r-by-n_t complex channel matrix as

$$\mathbf{H}(t) = \begin{bmatrix} h_{11}(t) & h_{21}(t) & \cdots & h_{n_t 1}(t) \\ h_{12}(t) & h_{22}(t) & \cdots & h_{n_t 2}(t) \\ \vdots & \vdots & \ddots & \vdots \\ h_{1n_r}(t) & h_{2n_r}(t) & \cdots & h_{n_t n_r}(t) \end{bmatrix} \tag{2.4}$$

$\mathbf{H}(t) \in \mathbb{C}^{n_r \times n_t}$ is the normalized channel matrix. Normalization is done such that each element of $\mathbf{H}$ has a spatial average power loss of unity. The normalized channel matrix $\mathbf{H}$ is drawn from the following independent and identically distributed (iid), complex, zero-mean and unit-variance entries:

$$h_{ij} \simeq \mathcal{N}\left(0, 1/\sqrt{2}\right) + \sqrt{-1}\mathcal{N}\left(0, 1/\sqrt{2}\right)$$

where $|H_{ij}|^2$ is a normalized chi-squared (χ_2^2) random variable such that, $\mathcal{E}[|h_{ij}|^2] = 1$. The channel is assumed to exhibit Rayleigh flat fading, and the channel matrix is represented by $n_r \times n_t$ complex values that are constant over the band of interest. Moreover, the average channel gain grows linearly with the number of receive antennas, that is, $\mathcal{E}\{\|\mathbf{h}_i\|^2\} = n_r$, where $\mathbf{h}_i$ is the ith column of channel matrix $\mathbf{H}$.

- The system of equations

$$y_i(t) = \sum_{k=1}^{n_t} h_{ik}(t)x_k(t) + v_i(t) \quad i = 1, 2, \ldots, n_r, \quad k = 1, 2, \ldots, n_t \quad (2.5)$$

defines the complex signal received at the ith antenna due to the transmitted symbol radiated by the kth antenna. The term $v_i(t)$ denotes the additive complex channel noise perturbing $y_i(t)$. Let the n_r-by-1 vector

$$\mathbf{y}(t) = [y_1(t), y_2(t), \ldots, y_{n_r}(t)]^T \quad (2.6)$$

denote the complex received signal vector, and let the n_r-by-1 vector

$$\mathbf{v}(t) = [v_1(t), v_2(t), \ldots, v_{n_r}(t)]^T \quad (2.7)$$

denote the complex channel noise vector. We may then rewrite the system of equations in the compact matrix form

$$\mathbf{y}(t) = \mathbf{H}(t)\mathbf{x}(t) + \mathbf{v}(t), \quad t = 1, 2, \ldots, N_s \quad (2.8)$$

Equation (2.8) describes the basic complex channel model for MIMO wireless communications, assuming the use of a flat-fading channel. The equation describes the input-output behavior of the channel at discrete time t. To simplify the exposition, hereafter we suppress the dependence on time t by writing

$$\mathbf{y} = \mathbf{Hx} + \mathbf{v} \quad (2.9)$$

where it is understood that all four vector/matrix terms of the equation $\mathbf{x}$, $\mathbf{H}$, $\mathbf{v}$, and $\mathbf{y}$ are in fact dependent on the discrete time t. Figure 2.2 depicts the basic channel model of (2.9).

When the channel is constant for at least N_s channel uses (representing a quasi-static scenario), we define $\mathbf{X} = [\mathbf{x}(1), \mathbf{x}(2), \ldots, \mathbf{x}(N_s)]$, $\mathbf{Y} = [\mathbf{y}(1), \mathbf{y}(2), \ldots, \mathbf{y}(N_s)]$, and $\mathbf{V} = [\mathbf{v}(1), \mathbf{v}(2), \ldots, \mathbf{v}(N_s)]$. We then write

$$\mathbf{Y} = \mathbf{HX} + \mathbf{V} \quad (2.10)$$

where $\mathbf{X} \in \mathbb{C}^{n_t \times N_s}$, $\mathbf{Y} \in \mathbb{C}^{n_r \times N_s}$, and $\mathbf{V} \in \mathbb{C}^{n_r \times N_s}$.

For mathematical tractability, we assume a Gaussian model made up of three elements relating to the transmitter, channel, and receiver, respectively:

1. The n_t symbols constituting the transmitted signal vector $\mathbf{x}$ are drawn from a white complex Gaussian codebook; that is, the symbols are i.i.d.

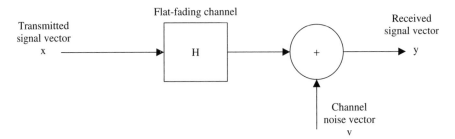

Figure 2.2 Depiction of the basic channel model of (2.9).

complex Gaussian random variables with zero mean and common variance σ_x^2. Hence, the correlation matrix of the transmitted signal vector $\mathbf{x}$ is defined by

$$\mathbf{R}_x = \mathcal{E}\left[\mathbf{x}\mathbf{x}^\dagger\right] \tag{2.11}$$

$$= \sigma_x^2 \mathbf{I}_{n_t} \tag{2.12}$$

where $\mathbf{I}$ is the n_t-by-n_t identity matrix.

2. The $n_t \times n_r$ elements of the channel matrix $\mathbf{H}$ are drawn from an ensemble of i.i.d. complex random variables with zero mean and unit variance, as shown by the complex distribution

$$h_{ij} \simeq \mathcal{N}\left(0, 1/\sqrt{2}\right) + \sqrt{-1}\mathcal{N}\left(0, 1/\sqrt{2}\right) \tag{2.13}$$

where $\mathcal{N}(.,.)$ denotes a real Gaussian distribution. On this basis, we find that the squared amplitude component, namely, $|h_{ik}|^2$, is a chi-square random variable with the mean

$$E\left[|h_{ij}|^2\right] = 1 \quad \text{for all} \quad i, j \tag{2.14}$$

3. The n_t elements of the channel noise vector w are i.i.d. complex Gaussian random variables with zero mean and common variance; that is, the correlation matrix of the noise vector $\mathbf{v}$ is given by

$$\mathbf{R}_v = E[\mathbf{v}\mathbf{v}^\dagger]$$
$$= \sigma_v^2 \mathbf{I}_{n_r} \tag{2.15}$$

where $\mathbf{I}_{n_r}$ is the n_r-by-n_r identity matrix.

The average signal-to-noise ratio (SNR) at each receiver input is given by, in light of (2.3) and the assumption that is a normalized random variable with zero mean and unit variance,

$$\rho = \frac{P}{\sigma_v^2}$$

$$= \frac{n_t \sigma_x^2}{\sigma_v^2}$$

(2.16)

which, for a prescribed noise variance σ_v^2, is fixed once the total transmit power P is fixed. Note also that (1) all the n_t transmitted signals occupy a common channel bandwidth and (2) the SNR ρ is independent of n_r.

The idealized Gaussian model described herein is applicable to indoor local area networks and other wireless environments, where user terminals' mobility is limited. The model, however, ignores the unavoidable ambient noise, which, as a result of experimental measurements, is known to be decidedly non-Gaussian due to the impulse nature of man-made electromagnetic interference as well as natural noise.

2.3 CHANNEL CAPACITY

The information capacity of a real additive white Gaussian noise (AWGN) channel, subject to the constraint of a fixed transmit power P, is defined by

$$C = B \cdot \log_2 \left(1 + \frac{P}{\sigma_v^2} \right) \text{ bits/s}$$

(2.17)

where B is the channel bandwidth and σ_v^2 is the noise variance measured over the bandwidth B.

Given a time-invariant channel, (2.17) defines the maximum data rate that can be transmitted over the channel with an arbitrarily small probability of error being incurred as a result of the transmission. With the channel used K times for the transmission of K symbols in T seconds, say, the transmission capacity per unit time is (K/T) times the formula for C given in (2.17). Recognizing that $K = 2BT$, in accordance with the sampling theorem, we may express the information capacity of the AWGN channel in the equivalent form

$$C = \frac{1}{2} \cdot \log_2 \left(1 + \frac{P}{\sigma_v^2} \right) \text{ bits/s/Hz}$$

(2.18)

Note that one bit per second per hertz corresponds to one bit per transmission.

With wireless communication as the medium of interest, consider next the case of a complex, flat-fading channel with the receiver having perfect knowledge of the channel state. The capacity of such a channel is given by

$$C = E \left[\log_2 \left(1 + \frac{|h|^2 P}{\sigma_v^2} \right) \right] \text{ bits/s/Hz}$$

(2.19)

where the expectation is taken over the gain of the channel, $h(t)$, and the channel is assumed to be stationary and ergodic. In recognition of this assumption, C is commonly referred to as the ergodic capacity of the flat-fading channel, and the channel coding is applied across fading intervals (i.e., over an "ergodic" interval of channel variation with time). It is important to note that the scaling factor of 1/2 is missing from the capacity formula of (2.19). The reason for this omission is that (2.19) refers to a complex baseband channel, whereas (2.17) refers to a real channel. The fading channel covered by (2.19) operates on a complex signal, namely, a signal with in-phase and quadrature components. Therefore, such a complex channel is equivalent to two real channels with equal capacities and operating in parallel, hence the result presented in (2.19).

Equation (2.19) applies to the simple case of a single-input, single-output (SISO) flat-fading channel. Generalizing this formula to the case of a MIMO flat-fading channel governed by the Gaussian model, we find that the ergodic capacity of the MIMO channel is given by

$$C = E\left[\log_2\left\{\frac{\det(\mathbf{R}_v + \mathbf{H}\mathbf{R}_x\mathbf{H}^{\dagger})}{\det(\mathbf{R}_v)}\right\}\right] \text{ bits/s/Hz} \qquad (2.20)$$

which is subject to the constraint $P = $ constant transmit power.

Next, we present the derivation of the log-det capacity formula of the MIMO link given in Eq. (2.20).

2.3.1 Information Theory in Complex Multidimensional Gaussian Distribution

The purpose of this section is to present some basic concepts in information theory on complex multidimensional Gaussian distributed signals that are needed in the derivation of the log-det capacity formula for MIMO channels. To prepare the way for the derivation, we briefly review some basic concepts in information theory that are needed.

Consider a continuous random-variable X with the probability density function $f_X(x)$. The differential entropy of the random variable X, measured in bits, is defined by

$$h(X) = -\int_{-\infty}^{\infty} f_X(x) \log_2(f_X(x))dx$$

$$= -E[\log_2 f_X(x)] \quad \text{bits} \qquad (2.21)$$

where E is the statistical expectation operator. It is important to note that the symbol X in the entropy $h(X)$ is not the argument of a function; rather, it serves as a label for the source of information.

When we have a continuous random vector $\mathbf{X}$ consisting of N random variables $X_1, X_2, \ldots, X_N$, we may generalize (2.21) and define the differential entropy of $\mathbf{X}$

as the N-fold integral

$$h(\mathbf{X}) = -\int_{-\infty}^{\infty} f_{\mathbf{X}}(\mathbf{x}) \log_2(f_{\mathbf{X}}(\mathbf{x}))d\mathbf{x}$$

$$= -E[\log_2 f_{\mathbf{X}}(\mathbf{x})] \quad \text{bits} \tag{2.22}$$

where $f_{\mathbf{X}}(\mathbf{x})$ is the joint probability density function of the random vector $\mathbf{X}$.

The logarithmic description of entropy is evident from both (2.21) and (2.22). This particular form of description is in perfect accord with the notion of entropy in thermodynamics.

Equations (2.21) and (2.22) apply to random data, whether real or complex. The difference between these two forms of data is in the way in which the pertinent probability density functions are defined, as illustrated in the following.

We consider an N-dimensional complex Gaussian-distributed vector $\mathbf{X}$. Each element of $\mathbf{X}$ consists of an in-phase component $X_{k,I}$ and a quadrature component $X_{k,Q}$, as shown by

$$X_k = X_{k,I} + jX_{k,Q} \quad k = 1, 2, \ldots, N \tag{2.23}$$

and

$$\mathbf{X} = \mathbf{X}_I + j\mathbf{X}_Q \tag{2.24}$$

It is assumed that $\mathbf{X}$ has zero mean. The requirement is to determine the differential entropy of $\mathbf{X}$. If the components $\mathbf{X}_I$ and $\mathbf{X}_Q$ are orthogonal, that is, if we have

$$E[\mathbf{X}_I\mathbf{X}_Q^T] = 0 \tag{2.25}$$

and they are both Gaussian distributed, then they are statistically independent, as shown by

$$f_{\mathbf{X}_I, \mathbf{X}_Q}(\mathbf{x}_I, \mathbf{x}_Q) = f_{\mathbf{X}_I}(\mathbf{x}_I) f_{\mathbf{X}_Q}(\mathbf{x}_Q) \tag{2.26}$$

The in-phase component $\mathbf{X}_I$ and the quadrature component $\mathbf{X}_Q$ share the same formula for their joint probability density functions. We therefore make two important observations:

1. The components $\mathbf{X}_I$ and $\mathbf{X}_Q$ have exactly the same entropy.
2. Since the differential entropy is logarithmic in nature, it follows that the differential entropies of $\mathbf{X}_I$ and $\mathbf{X}_Q$ are additive in calculating the differential entropy of $\mathbf{X}$.

Hence, we may write

$$h(\mathbf{X}_I) = h(\mathbf{X}_Q) \tag{2.27}$$

and

$$h(\mathbf{X}) = h(\mathbf{X}_I) + h(\mathbf{X}_Q)$$
$$= 2h(\mathbf{X}_I)$$

The joint probability density function of the complex Gaussian vector $\mathbf{X}$, with zero mean and correlation matrix $\mathbf{R_x}$, is defined by

$$f_{\mathbf{X}}(\mathbf{x}) = \frac{1}{(2\pi)^N \det(\mathbf{R_x})} \exp\left(-\frac{1}{2} \mathbf{x}^T R_{\mathbf{x}}^{-1} \mathbf{x} \right) \tag{2.28}$$

where $\mathbf{R_x}^{-1}$ is the inverse of $\mathbf{R_x}$ and $\det(\mathbf{R_x})$ is its determinant. Substituting (2.28) into (2.22), using the fact that the volume under $f_{\mathbf{X}}(\mathbf{x})$ is unity, and then simplifying terms, we get

$$h(\mathbf{X}) = N + N \log_2(2\pi) + \log_2 \det(\mathbf{R_x}) \text{ bits} \tag{2.29}$$

which is uniquely defined by the correlation matrix $\mathbf{R_x}$.

For the special case of a scalar complex Gaussian random variable X, $N = 1$ and (2.29) reduces to

$$h(X) = 1 + \log_2\left(2\pi\sigma_X^2\right) \text{ bits} \quad X : \text{complex} \tag{2.30}$$

where σ_X^2 is the variance of X. If X is real, we have

$$h(X) = \frac{1}{2}\left[1 + \log_2\left(2\pi\sigma_X^2\right)\right] \text{ bits} \qquad (X : \text{real}) \tag{2.31}$$

For a given variance σ_X^2, the Gaussian random variable X has the largest differential entropy attainable by any random variable in its class (i.e., real or complex). The same is true of a multivariate Gaussian distribution.

For the discussion at hand, we need one other notion: mutual information, which applies to a pair of related random variables or random vectors. To be specific, consider a pair of random variables X and Y with joint probability density function $f_{X,Y}(x, y)$. The mutual information between X and Y is defined by

$$I(X; Y) = \int_{-\infty}^{\infty} \int_{-\infty}^{\infty} f_{X,Y}(x, y) \log_2\left(\frac{f_{X|Y}(x|y)}{f_X(x)} \right) dx\, dy \tag{2.32}$$

where $f_{X|Y}(x|y)$ is the conditional probability density function of X, given $Y = y$. In other words, the mutual information $I(X; Y)$ is a measure of the uncertainty about the random variable X that is resolved by observing the second random variable Y.

On the basis of (2.32), we may derive the following properties of mutual information that hold in general:

$$I(X; Y) \geq 0 \tag{2.33}$$

$$I(X; Y) = I(Y; X) \tag{2.34}$$

$$I(X; Y) = h(X) - h(X|Y)$$
$$= h(Y) - h(Y|X) \tag{2.35}$$

where $h(X)$ and $h(Y)$ are the differential entropies of X and Y, respectively, and $h(X|Y)$ is the conditional differential entropy of X, given Y, as shown by

$$h(X|Y) = - \int_{-\infty}^{\infty} \int_{-\infty}^{\infty} f_{X,Y}(x, y) \log_2 f_{X|Y}(x|y) dx dy \tag{2.36}$$

Formulas similar to (2.32) through (2.36) apply to a related pair of random vectors $\mathbf{X}$ and $\mathbf{Y}$.

With the definitions of differential entropy, conditional differential entropy, and mutual information at hand, we are ready to proceed with the derivation of the log-det-capacity formula, as described next.

2.4 MIMO CAPACITY FOR A CHANNEL KNOWN AT THE RECEIVER

With the basic complex channel model and the preliminaries of information theory concepts applied to complex multidimensional Gaussian distributed signals at hand, we are now ready to focus the discussion on the primary issue of interest: the channel capacity of a MIMO wireless link. Two cases will be considered in what follows. The first case, described in Sections 2.4.1 and 2.4.2, considers a link that is stationary and therefore ergodic. The second case, described in Section 2.4.3, considers a nonergodic link, assuming quasi-stationarity from one data burst to another.

2.4.1 Ergodic Capacity

Consider a communication link with multiple antennas, with the n_t-by-1 vector $\mathbf{x}$ denoting the transmitted signal vector and the n_r-by-1 vector $\mathbf{y}$ denoting the received signal vector. These two vectors are related by the input-output relation of the channel:

$$\mathbf{y} = \mathbf{Hx} + \mathbf{v} \tag{2.37}$$

where $\mathbf{H}$ is the channel matrix of the link and $\mathbf{v}$ is the additive channel noise vector. The vectors $\mathbf{x}$, $\mathbf{v}$, and $\mathbf{y}$ are realizations of the random vectors X, V, and Y.

In this section, the following assumptions are made:

1. The channel is stationary and ergodic.
2. The channel matrix $\mathbf{H}$ is made up of i.i.d. Gaussian elements.

3. The channel state $\mathbf{H}$ is known to the receiver but not the transmitter.
4. The transmitted signal vector $\mathbf{x}$ has zero mean and correlation matrix $\mathbf{R_x}$.
5. The additive channel noise vector $\mathbf{v}$ has zero mean and correlation matrix $\mathbf{R_v}$.
6. Both $\mathbf{x}$ and $\mathbf{v}$ are governed by Gaussian distributions.

With both $\mathbf{H}$ and $\mathbf{x}$ unknown to the transmitter, the primary issue is to determine $I(\mathbf{x;y,H})$ that denotes the mutual information between the transmitted signal vector $\mathbf{x}$ and both the received signal vector $\mathbf{y}$ and the channel matrix $\mathbf{H}$. Extending the definition of mutual information given in (2.32) to the problem at hand, we write

$$I(\mathbf{X}; \mathbf{Y}, \mathbf{H}) = \int_{\mathcal{H}} \int_{\mathcal{S}} \int_{\mathcal{X}} f_{\mathbf{X},\mathbf{Y},\mathbf{H}}(\mathbf{x}, \mathbf{y}, \mathbf{H}) \log_2 \left(\frac{f_{\mathbf{X}|\mathbf{Y},\mathbf{H}}(\mathbf{x}|\mathbf{y}, \mathbf{H})}{f_{\mathbf{Y},\mathbf{H}}(\mathbf{y}, \mathbf{H})} \right) d\mathbf{x}d\mathbf{y}d\mathbf{H} \quad (2.38)$$

where $\mathcal{X}$, $\mathcal{Y}$, and $\mathcal{H}$ are the respective spaces pertaining to the random vectors $\mathbf{X}$ and $\mathbf{Y}$ and the matrix $\mathbf{H}$. According to probability theory, we have

$$f_{\mathbf{X},\mathbf{Y},\mathbf{H}}(\mathbf{x}, \mathbf{y}, \mathbf{H}) = f_{\mathbf{x},\mathbf{Y}|\mathbf{H}}(\mathbf{x}, \mathbf{y}|\mathbf{H}) f_{\mathbf{H}}(\mathbf{H}) \quad (2.39)$$

We may therefore rewrite (2.38) in the equivalent form

$$I(\mathbf{X}; \mathbf{Y}, \mathbf{H}) = \int_{\mathcal{H}} f_{\mathbf{H}}(\mathbf{H}) \left[\int_{\mathcal{Y}} \int_{\mathcal{X}} f_{\mathbf{X},\mathbf{Y}|\mathbf{H}}(\mathbf{x}, \mathbf{y}|\mathbf{H}) \log_2 \left(\frac{f_{\mathbf{X}|\mathbf{Y},\mathbf{H}}(\mathbf{x}|\mathbf{y}, \mathbf{H})}{f_{\mathbf{Y},\mathbf{H}}(\mathbf{y}, \mathbf{H})} \right) d\mathbf{x}d\mathbf{y} \right] d\mathbf{H}$$

$$= E_{\mathbf{H}} \left[\int_{\mathcal{Y}} \int_{\mathcal{X}} f_{\mathbf{X},\mathbf{Y}|\mathbf{H}}(\mathbf{x}, \mathbf{y}|\mathbf{H}) \log_2 \left(\frac{f_{\mathbf{X}|\mathbf{Y},\mathbf{H}}(\mathbf{x}|\mathbf{y}, \mathbf{H})}{f_{\mathbf{Y},\mathbf{H}}(\mathbf{y}, \mathbf{H})} \right) d\mathbf{x}d\mathbf{y} \right] d\mathbf{H} \quad (2.40)$$

$$= E_{\mathbf{H}} \left[I(\mathbf{x}; \mathbf{y}|\mathbf{H}) \right] \quad (2.41)$$

where the expectation is with respect to the channel matrix $\mathbf{H}$, and

$$I(\mathbf{x}; \mathbf{y}, \mathbf{H}) = \int_{\mathcal{Y}} \int_{\mathcal{X}} f_{\mathbf{X},\mathbf{Y}|\mathbf{H}}(\mathbf{x}, \mathbf{y}|\mathbf{H}) \log_2 \left(\frac{f_{\mathbf{X}|\mathbf{Y},\mathbf{H}}(\mathbf{x}|\mathbf{y}, \mathbf{H})}{f_{\mathbf{Y},\mathbf{H}}(\mathbf{y}, \mathbf{H})} \right) d\mathbf{x}d\mathbf{y} \quad (2.42)$$

is the conditional mutual information between the transmitted signal vector $\mathbf{x}$ and the received signal vector $\mathbf{y}$, given the channel matrix $\mathbf{H}$. However, by assumption, the channel state is unknown to the receiver. It follows therefore that as far as the receiver is concerned, $I(\mathbf{x}; \mathbf{y}|\mathbf{H})$ is a random variable, hence the expectation with respect to $\mathbf{H}$ in (2.41). The quantity resulting from this expectation is deterministic, defining the mutual information jointly between the transmitted signal vector $\mathbf{x}$ and both the received signal vector $\mathbf{y}$ and the channel matrix $\mathbf{H}$. The result so obtained is indeed consistent with what we know about the notion of joint mutual information. Next, applying the vector form of the first line in (2.35) to the mutual information $I(\mathbf{x}; \mathbf{y}|\mathbf{H})$, we may write

$$I(\mathbf{x}; \mathbf{y}|\mathbf{H}) = h(\mathbf{y}|\mathbf{H}) - h(\mathbf{y}|\mathbf{x}, \mathbf{H}) \quad (2.43)$$

where $h(\mathbf{y}|\mathbf{H})$ is the conditional differential entropy of the input $\mathbf{x}$, given $\mathbf{H}$, and $h(\mathbf{y}|\mathbf{x}, \mathbf{H})$ is the conditional differential entropy of the output $\mathbf{y}$ given both $\mathbf{x}$ and $\mathbf{H}$. Both of these entropies are random quantities, as they depend on $\mathbf{H}$ that is unknown to the receiver.

We now invoke the assumed Gaussianity of both $\mathbf{x}$ and $\mathbf{H}$, in which case $\mathbf{y}$ also assumes a Gaussian description. Under these assumptions, we may express the entropy of the received signal $\mathbf{y}$ of dimension n_r, given $\mathbf{H}$, as

$$h(\mathbf{y}|\mathbf{H}) = n_r + n_r \log_2(2\pi) + \log_2\{\det(\mathbf{R_y})\} \quad \text{bits} \tag{2.44}$$

where $\mathbf{R_y}$ is the correlation matrix of $\mathbf{y}$ Recognizing that the transmitted signal vector $\mathbf{x}$ and the channel noise vector $\mathbf{v}$ are independent of each other, we find from (2.37) that the correlation matrix of the received signal vector $\mathbf{y}$ is given by

$$
\begin{aligned}
\mathbf{R_y} &= E[\mathbf{yy}^\dagger] \\
&= E[(\mathbf{Hx} + \mathbf{v})(\mathbf{Hx} + \mathbf{v})^\dagger] \\
&= E[(\mathbf{Hx} + \mathbf{v})(\mathbf{x}^\dagger\mathbf{H}^\dagger + \mathbf{v}^\dagger)] \\
&= E[\mathbf{Hxx}^\dagger\mathbf{H}^\dagger] + E[\mathbf{vv}^\dagger] \quad \text{because } E[\mathbf{xv}^\dagger] = 0 \\
&= HE[\mathbf{xx}^\dagger]\mathbf{H}^\dagger + \mathbf{R_v} \\
&= \mathbf{HR_xH}^\dagger + \mathbf{R_v}
\end{aligned}
\tag{2.45}
$$

where

$$\mathbf{R_x} = E[\mathbf{xx}^\dagger] \tag{2.46}$$

and

$$\mathbf{R_v} = E[\mathbf{vv}^\dagger] \tag{2.47}$$

Hence, using (2.46) in (2.44), we get

$$h(\mathbf{y}|\mathbf{H}) = n_r + n_r \log_2(2\pi) + \log_2\{\det(\mathbf{R_v} + \mathbf{HR_xH}^\dagger)\} \quad \text{bits} \tag{2.48}$$

Next, we note that since the vectors $\mathbf{x}$ and $\mathbf{v}$ are independent and the sum of $\mathbf{v}$ plus $\mathbf{Hx}$ equals $\mathbf{y}$, as indicated in (2.37), the conditional differential entropy of $\mathbf{y}$, given both $\mathbf{x}$ and $\mathbf{H}$, is simply equal to the differential entropy of the additive channel noise vector $\mathbf{v}$, as shown by

$$h(\mathbf{y}|\mathbf{x}, \mathbf{H}) = h(\mathbf{v}) \tag{2.49}$$

Again, invoking the formula of (2.29), we have

$$h(\mathbf{v}) = n_r + n_r \log_2(2\pi) + \log_2\{\det(\mathbf{R_v})\} \quad \text{bits} \tag{2.50}$$

Using (2.48), (2.49), and (2.50) in (2.43), we get

$$I(\mathbf{x}; \mathbf{y}|\mathbf{H}) = \log_2\left\{\det\left(\mathbf{R_v} + \mathbf{HR_xH}^H\right)\right\} - \log_2\{\det(\mathbf{R_v})\}$$

$$= \log_2\left\{\frac{\det\left(\mathbf{R_v} + \mathbf{HR_xH}^H\right)}{\det(\mathbf{R_v})}\right\}$$

(2.51)

As noted previously, the conditional mutual information $I(\mathbf{x}; \mathbf{y}|\mathbf{H})$ is a random variable. Hence, using (2.51) in (2.41), we finally formulate the ergodic capacity of the MIMO link as the expectation

$$C = E_{\mathbf{H}}\left[\log_2\left\{\frac{\det\left(\mathbf{R_v} + \mathbf{HR_xH}^H\right)}{\det(\mathbf{R_v})}\right\}\right] \quad \text{bits/s/Hz}$$

(2.52)

which is subject to the constraint

$$\max_{\mathbf{R}_x} \text{tr}[\mathbf{R_x}] \leq P \quad P = \text{constant transmitted power}$$

(2.53)

Equation (2.52) is the desired log-det formula for the ergodic capacity of the MIMO link. The expectation in (2.52) is over the random channel matrix $\mathbf{H}$, and the superscript † denotes Hermitian transposition; $\mathbf{R_x}$ and $\mathbf{R_v}$ are, respectively, the correlation matrices of the transmitted signal vector $\mathbf{x}$ and the channel noise vector $\mathbf{v}$. This formula is of general applicability in that correlations among the elements of the transmitted signal vector $\mathbf{x}$ and among those of the channel noise vector $\mathbf{w}$ are permitted. However, the assumptions made in its derivation involve the Gaussianity of $\mathbf{x}$, $\mathbf{H}$ and $\mathbf{v}$.

In general, it is difficult to evaluate (2.52) except for the Gaussian model. In particular, substituting (2.12) and (2.16) in (2.52) yields, after simplification,

$$C = E\left[\log_2\left\{\det\left(\mathbf{I}_{n_r} + \frac{\sigma_x^2}{\sigma_v^2}\mathbf{HH}^\dagger\right)\right\}\right] \quad \text{bits/s/Hz}$$

(2.54)

Invoking the definition of the average SNR ρ introduced in (2.17), we may rewrite (2.54) in the equivalent form

$$C = E\left[\log_2\left\{\det\left(\mathbf{I}_{n_r} + \frac{\rho}{n_t}\mathbf{HH}^\dagger\right)\right\}\right] \quad \text{bits/s/Hz}$$

(2.55)

The formula of (2.55), defining the ergodic capacity of a MIMO flat-fading channel, involves the determinant of an n_r-by-n_r sum matrix followed by the logarithm to base 2. It is for this reason that we refer to this formula as the log-det capacity formula for a Gaussian MIMO channel.

Clearly, the capacity formula of (2.19), pertaining to a conventional flat-fading link with a single antenna at both ends of the link, is a special case of the log-det capacity formula. Specifically, for $n_t = n_r = 1$ (i.e., no spatial diversity) and $\mathbf{H} = h$ (with dependence on discrete-time n), (2.54) reduces to (2.19).

Another insightful result that follows from the log-det capacity formula is that if $n_t = n_r = N$, then as N approaches infinity, the capacity C defined in (2.54) grows asymptotically (at least) linearly with N, as shown by

$$\lim_{N \to \infty} \frac{C}{N} \geq \text{constant} \tag{2.56}$$

In words, the asymptotic formula of (2.56) may be stated as follows: The ergodic capacity of a MIMO flat-fading wireless link with an equal number of transmit and receive antennas, N, grows roughly proportionately with N

What this statement teaches us is that by increasing the computational complexity resulting from the use of multiple antennas at both the transmit and receive ends of a wireless link, we are able to increase the spectral efficiency of the link far more than is possible by conventional means (e.g., increasing the transmit SNR). The potential for this very sizable increase in the spectral efficiency of a MIMO wireless communication system is attributed to the key parameter $\min\{n_t, n_r\}$, which defines the number of degrees of freedom per second per Hertz provided by the system.

Moreover, at high SNRs, the capacity gain of a MIMO wireless communication with the channel state known to the receiver is $\min\{n_t, n_r\}$, bits per second per hertz for every 3 dB increase in the SNR.

One last comment is in order. The white Gaussian input spectrum

$$\mathbf{R_x} = \sigma_x^2 \mathbf{I}_{n_t} \tag{2.57}$$

is not necessarily optimal; nevertheless, its application does yield a lower bound on the ergodic capacity C.

2.4.2 Two Other Special Cases of the Log-Det Formula: Capacities of Receive and Transmit Diversity Links

Naturally, the log-det capacity formula of (2.55) for the channel capacity of a (n_t, n_r) wireless link includes the channel capacities of receive and transmit diversity links as special cases, as shown next.

- Diversity-on-receive channel. For $n_t = 1$, H reduces to a row vector, and with it Eq. (2.55) reduces to

$$C = E \left[\log_2 \left\{ \left(1 + \rho \sum_{i=1}^{n_r} |h_i|^2 \right) \right\} \right] \quad \text{bits/s/Hz} \tag{2.58}$$

where, compared to the standard channel capacity of (2.19), the squared Euclidean norm is replaced by the sum of squared amplitudes $|h_i|^2$, $i = 1, 2, \ldots, n_r$. Equation (2.58) expresses the ergodic capacity due to the linear combination of the receive-antenna outputs, which is designed to maximize the information contained in the n_r received signals about the transmitted signal. This is simply a restatement of the maximal-ratio combining principle.

- Diversity-on-transmit channel. For $n_r = 1$, (2.55) reduces to

$$C = E\left[\log_2\left(1 + \frac{\rho}{n_t} \sum_{k=1}^{n_t} |h_k|^2 \right) \right] \text{ bits/s/Hz} \qquad (2.59)$$

where the simple squared norm is again replaced by the sum of squared amplitudes $|h_k|^2$, $k = 1, 2, \ldots, n_t$. Compared to Case 1 on receive diversity, the channel capacity of diversity-on-transmit is reduced because the total transmit power is held constant, independently of the number of transmit antennas, n_t.

2.4.3 Outage Capacity

To realize the log-det capacity formula of (2.55), the MIMO channel code needs to see an ergodic process of the random channel processes, which, in turn, results in a hardening of the rate of reliable transmission to the $E[\log 2\det()]$ information rate. As in all information-theoretic arguments, the bit error rate would go to zero asymptotically in the block length of the code, thereby entailing a long transportation delay from the sender to the sink. In practice, however, the MIMO wireless channel is often nonergodic, and the requirement is to operate the channel under delay constraints. The issue of interest is then summed up as follows:

> How much information can be transmitted across a nonergodic channel, particularly if the channel code is long enough to see just one random channel matrix?

In the situation described here, the rate of reliable information transmission (i.e., the strict Shannon-sense capacity) is zero since for any positive rate, there exists a nonzero probability that the channel would not support such a rate.

To get around this serious difficulty, the notion of outage is introduced into the characterization of the MIMO link. Specifically, the outage probability of a MIMO link is defined as the probability for which the link is in a state of outage (i.e., failure) for data transmitted across the link at a certain rate R, where R is measured in bits per second per hertz. To proceed on this probabilistic basis, it is customary to operate the MIMO link by transmitting data in the form of bursts or frames and to invoke a quasi-static model governed by four conditions:

1. The burst is long enough to accommodate the transmission of a large number of symbols, which, in turn, permits the use of an idealized infinite-time horizon basic to information theory.
2. Yet, the burst is short enough to treat the wireless link as quasi-static during each burst; the slow variation is used to justify the assumption that the receiver has perfect knowledge of the channel state.
3. The channel matrix is permitted to change from burst k, say, to the next one, $k + 1$, thereby accounting for statistical variations of the link.

4. The different realizations of the transmitted signal vector $\mathbf{x}$ are drawn from a white Gaussian codebook; that is, the correlation matrix of s is defined by (2.12).

The quasi-stationary model is applicable to outdoor wireless environments characterized by highly mobile user terminals, which makes the environment nonstationary and therefore nonergodic.

In any event, conditions 1 and 4 pertain to signal transmission, while conditions 2 and 3 pertain to the channel.

To proceed with the evaluation of outage probability, we first note that conditions 1 through 4 of the stochastic model just described for a nonstationary wireless link permit us to build on some of the results discussed in Subsection 2.4.1. In particular, in light of the log-det capacity formula of (2.55), we may view the random variable

$$C = \log_2 \left\{ \det\left(\mathbf{I}_{n_r} + \frac{\rho}{n_t} \mathbf{H}_k \mathbf{H}_k^\dagger \right) \right\} \quad \text{bits/s/Hz} \quad \text{for burst } k \qquad (2.60)$$

as the expression for capacity point k of the cumulative probability distribution function of the wireless link. In other words, with the random channel matrix $\mathbf{H}_k$ varying from one burst to the next, C_k will itself vary in a corresponding way. A consequence of this random behavior is that occasionally a draw from the cumulative distribution function of the wireless link results in an inadequate value for C_k to support reliable communication over the link, in which case the link is said to be in an outage state. Correspondingly, for a given transmission strategy, we define the outage probability at rate R as

$$P_{\text{outage}} = \Pr\left\{ C_k < R \text{ for some burst } k \right\} \qquad (2.61)$$

or, equivalently,

$$P_{\text{outage}} = \Pr\left\{ \log_2 \left\{ \det\left(\mathbf{I}_{n_r} + \frac{\rho}{n_t} \mathbf{H}_k \mathbf{H}_k^\dagger \right) \right\} < R \text{ for some burst } k \right\} \qquad (2.62)$$

On this basis, we may define the outage capacity of the MIMO link as the maximum bit rate that can be maintained across the link for all bursts of data transmissions (i.e., all possible channel states) for a prescribed outage probability.

To calculate the outage probability, we use the complementary cumulative distribution function of the random channel matrix $\mathbf{H}$ rather than the cumulative probability function itself. By definition, the complementary cumulative distribution function is equal to unity minus the cumulative distribution function.

Example 1 *In this example, we show the outage capacity for different antenna configurations and varying SNRs.*

In light of the random nature of the channel transfer function **H**, the outage capacity is evaluated using Monte Carlo simulation by computing the cumulative distribution function of the wireless link for a large number of statistically different realizations of **H**. To illustrate the simulation procedure, suppose we wish to calculate the outage capacity $C_{15}\%$ for error-free transmission for $100 - 15 = 85\%$ of the time. The calculation is performed for a (2,2) antenna configuration operating at an SNR of 10 dB, that is, $\rho = 10$.

We first obtain the cumulative distribution function for this wireless link, which is accomplished by generating a large number of Rayleigh-distributed random transfer functions under the flat-fading assumption. According to (2.60), the capacity is given by

$$C = \log_2 \left\{ \det\left(\mathbf{I}_2 + \frac{10}{2}\mathbf{H}_k\mathbf{H}_k^\dagger\right) \right\} \quad \text{bits/s/Hz}$$

for realization k of the channel transfer function. Figure 2.3(a) plots the histogram (i.e., probability density function) of the resulting channel capacity data.

Integrating the probability density function curve of Figure 2.3(a) and then subtracting the result from unity yields the complementary cumulative distribution function of the link, which is plotted in Figure 2.3(b). By definition, a plot of the complementary cumulative distribution function reveals the probability that a sample capacity will be greater than a threshold value (i.e., outage capacity) at a particular SNR (10 dB in this example) and for a specific antenna configuration (2-by-2 in this example). Thus, from Figure 2.3(b), we see the probability the channel maintains a capacity of 4.2 bits/s/Hz for 85% of the time.

Figure 2.4 displays two different plots of the outage capacity for different SNRs and antenna configurations. Part (a) of the figure plots the probability that the channel capacity is greater than the abscissa an versus the capacity in bits/s/Hz for a signal-to-noise ratio of $\rho = 10$ dB. Part (b) of the figure portrays the picture differently by plotting the outage capacity in bits/s/Hz versus the SNR in decibels.

Figure 2.4 clearly demonstrates two important points:

1. For $n_t = 1$, increasing the number of receive antennas by using $n_r = 1, 2, 4$, produces a modest increase in the outage capacity for a fixed SNR.
2. For $n_t = n_r$ and a fixed SNR, there is a significant increase in the outage capacity in going from antenna configuration (4,1) to (4,2), and a much bigger increase in the outage capacity in going from antenna configuration (4,2) to (4,4).

Example 2 *The previous example demonstrated the outage capacity of various diversity schemes.*

In this example, we consider a system with equal numbers of transmit and receive antennas (N) to demonstrate the multiplexing gain of MIMO systems. Figure 2.5 depicts the capacity complementary cumulative distribution function for $N = 1, 2, 4, 8$ and $N = 16$ at SNR = 18 dB. The figure reveals the significant

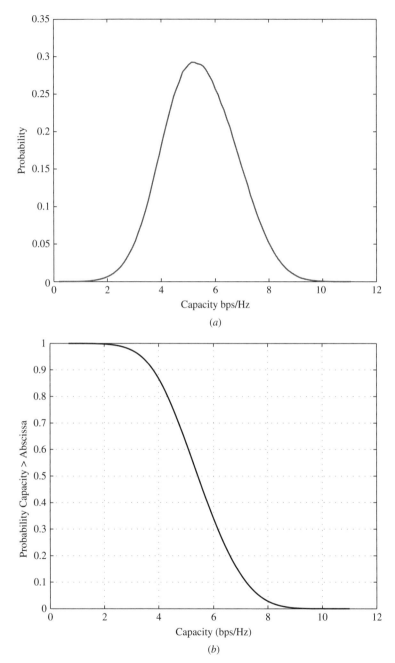

Figure 2.3 (*a*) Histogram (probability density function) of channel data for the SNR $\rho = 10\,\text{dB}$. (*b*) Complementary cumulative probability distribution function corresponding to the histogram of part (*a*).

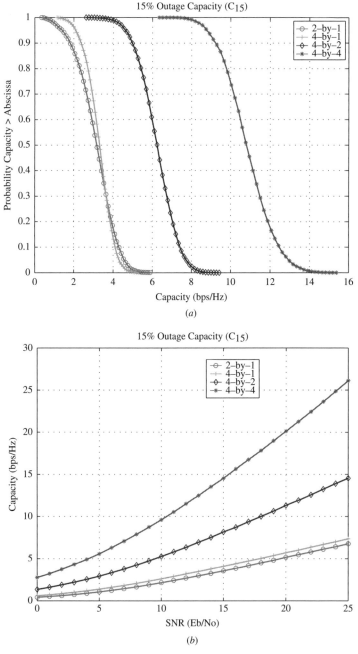

Figure 2.4 (a) Plots of the probability that the channel capacity is greater than the abscissa for five antenna configurations: $n_t = 1, n_r = 1$; $n_t = 4, n_r = 1$; $n_t = n_r = 2$; $n_t = 4, n_r = 2$; $n_t = 4, n_r = 4$; (b) plots of the outage capacity versus the SNR for the five antenna configurations given in part (a).

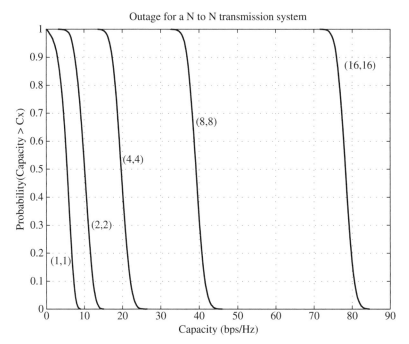

Figure 2.5 Outage capacity of an (N, N) system with $N = 1$, 2, 4, 8, and 16, where $n_t = n_r = N$.

capacity increment, at different outage probability tails, by doubling the number of transmit and receive antennas.

Figure 2.5 clearly demonstrates the power of multiplexing gain by increasing the size of a MIMO system. Note that Eq. (2.56) holds even for outage capacity of quasi-fading channels; that is, for large values of N, the random capacities converge to a fixed value.

It is of interest to note that at SNRs, the outage probability $P_{outage}(R)$ as defined in (2.62) is approximately the same as the frame (burst)-error probability in terms of the SNR exponent. Accordingly, we may use an analysis based on the outage probability to evaluate the performance of practical space-time block coding techniques. That is, for a prescribed rate R, we may evaluate how the performance of a certain space-time block coding technique compares to that predicted through an outage analysis or measurement.

2.5   CHANNEL KNOWN AT THE TRANSMITTER

The log-det capacity formula of (2.55) is based on the premise that the transmitter has no knowledge of the channel state. Knowledge of the channel state can be made available to the transmitter by first estimating the channel matrix **H** at the receiver and then sending this estimate to the transmitter via a feedback channel. In this scenario, the capacity is optimized over the correlation matrix of the transmitted signal vector **x**, subject to the power constraint, namely, that the trace of this correlation matrix is less than or equal to the constant transmit power P. It is important to note, however, that the capacity gain provided by knowledge of the channel state at the transmitter over the log-det formula of (2.55) is significant only at low SNRs; the gain reduces to zero as the signal-to-noise ratio increases.

To begin the exposition, consider the matrix product $\mathbf{HH}^\dagger$ in the log-det capacity formula of (2.55). The matrix product $\mathbf{HH}^\dagger$ satisfies the Hermitian property for all **H**. We may therefore diagonalize $\mathbf{HH}^\dagger$ by invoking the eigendecomposition of a Hermitian matrix and so write

$$\mathbf{U}^\dagger\mathbf{HH}^\dagger\mathbf{U} = \Lambda \tag{2.63}$$

where the two new matrices **U** and $\mathbf{HH}^\dagger$ are described as follows:

- The matrix *mathbfU* is a diagonal matrix whose n_r elements are the eigenvalues of the matrix product $\mathbf{HH}^\dagger$.
- The matrix **U** is a unitary matrix whose n_r columns are the eigenvectors associated with the eigenvalues of $\mathbf{HH}^\dagger$.

By definition, the inverse of a unitary matrix is equal to the Hermitian transpose of the matrix, as shown by

$$\mathbf{U}^{-1} = \mathbf{U}^\dagger \tag{2.64}$$

or, equivalently,

$$\mathbf{UU}^\dagger = \mathbf{U}^\dagger\mathbf{U} = \mathbf{I}_{n_r} \tag{2.65}$$

where is the n_r-by-n_r identity matrix.

Let the n_t-by-n_t matrix **V** denote another unitary matrix, as shown by

$$\mathbf{VV}^\dagger = \mathbf{V}^\dagger\mathbf{V} = \mathbf{I}_{n_t} \tag{2.66}$$

where $\mathbf{I}_{n_t}$ is the n_t-by-n_t identity matrix. Since the multiplication of a matrix by the identity matrix leaves the matrix unchanged, we may inject the matrix product $\mathbf{VV}^\dagger$ into the center of the left-hand side of Eq. (2.63), as shown by

$$\mathbf{U}^\dagger\mathbf{H}(\mathbf{VV}^\dagger)\mathbf{H}^\dagger\mathbf{U} = \Lambda \tag{2.67}$$

The left-hand side of (2.67), representing a square matrix, is recognized as the product of two rectangular matrices: the n_r-by-n_t matrix $\mathbf{U}^\dagger\mathbf{HV}$, and the n_t-by-n_r matrix $\mathbf{V}^\dagger\mathbf{HU}$, which are the Hermitians of each other. Let the n_t-by-n_t matrix **D** denote a new diagonal matrix related to the n_r-by-n_r diagonal matrix with $n_r < n_t$

as follows:

$$\Lambda = [\mathbf{D}\ \mathbf{0}][\mathbf{D}\ \mathbf{0}]^{\dagger} \tag{2.68}$$

where the null matrix $\mathbf{0}$ is added to maintain proper overall dimensionality of the equation. Except for some zero elements, $\mathbf{D}$ is the square root of Λ. Then, examining (2.67) and (2.68) and comparing terms, we deduce the new decomposition

$$\mathbf{U}^{\dagger}\mathbf{H}\mathbf{V} = [\mathbf{D}\ \mathbf{0}] \tag{2.69}$$

Equation (2.69) is a mathematical statement of the singular value decomposition theorem, according to which we have the following descriptions:

- The elements of the diagonal matrix

$$\mathbf{D} = \text{diag}(d_1, d_2, \ldots, d_{n_t}) \tag{2.70}$$

  are the singular values of the channel matrix $\mathbf{H}$.
- The columns of the unitary matrix

$$\mathbf{U} = \text{diag}(\mathbf{u}_1, \mathbf{u}_2, \ldots, \mathbf{u}_{N_r}) \tag{2.71}$$

  are the left singular vectors of matrix $\mathbf{H}$.
- The columns of the second unitary matrix

$$\mathbf{V} = \text{diag}(\mathbf{v}_1, \mathbf{v}_2, \ldots, \mathbf{v}_{n_t}) \tag{2.72}$$

  are the right singular vectors of matrix $\mathbf{H}$.

Applying the singular value decomposition of (2.69) to the basic channel model of (2.10), we can show the following:

$$\bar{\mathbf{y}} = \mathbf{D}\bar{\mathbf{x}} + \bar{\mathbf{v}} \tag{2.73}$$

where

$$\bar{\mathbf{y}} = \mathbf{U}^{\dagger}\mathbf{y} \tag{2.74}$$

$$\bar{\mathbf{x}} = \mathbf{V}^{\dagger}\mathbf{x} \tag{2.75}$$

and

$$\bar{\mathbf{v}} = \mathbf{U}^{\dagger}\mathbf{v} \tag{2.76}$$

Using the definitions of (2.70) through (2.72), we may rewrite the decomposed channel model of (2.73) in the scalar form

$$\bar{y}_i = d_i\bar{x}_i + \bar{v}_i \quad i = [1, 2, \ldots, n_r] \quad i = 1, 2, \ldots, nr \tag{2.77}$$

According to (2.77), singular value decomposition of the channel matrix $\mathbf{H}$ has transformed the MIMO wireless link with $n_r < n_t$ into n_r virtual channels. The virtual channels are all decoupled from each other in that they constitute a parallel set of n_r SISO channels, with each virtual channel being described by the scalar input-output relation of (2.77).

2.5.1 Eigendecomposition of the Log-Det Capacity Formula

The log-det formula of (2.78) for the ergodic capacity of a MIMO link involves the matrix product $\mathbf{HH}^\dagger$. Using (2.64) in (2.63) leads to the spectral decomposition of $\mathbf{HH}^\dagger$ in terms of n_r eigenmodes, with each eigenmode corresponding to virtual data transmission using a pair of right- and left-singular vectors of the channel matrix $\mathbf{H}$ as the transmit and receive antenna weights, respectively. Thus, we may write

$$\mathbf{HH}^\dagger = \mathbf{U\Lambda U}^\dagger$$

$$= \sum_{i=1}^{N_r} \lambda_i \mathbf{u}_i \mathbf{u}_i^\dagger$$

(2.78)

where the outer product is an n_r-by-n_r matrix with a rank equal to 1. Moreover, substituting the first line of this decomposition into the determinant part of (2.55) yields

$$\det\left(\mathbf{I}_{N_r} + \frac{\rho}{n_t}\mathbf{HH}^\dagger\right) = \det\left(\mathbf{I}_{N_r} + \frac{\rho}{n_t}\mathbf{U\Lambda U}^\dagger\right)$$

(2.79)

Next, invoking the determinant identity

$$\det(\mathbf{I} + \mathbf{AB}) = \det(\mathbf{I} + \mathbf{BA})$$

(2.80)

and then using the defining (2.65), we may rewrite (2.79) in the equivalent form

$$\det\left(\mathbf{I}_{N_r} + \frac{\rho}{n_t}\mathbf{HH}^\dagger\right) = \det\left(\mathbf{I}_{N_r} + \frac{\rho}{n_t}\mathbf{UU}^\dagger\mathbf{\Lambda}\right)$$

$$= \det\left(\mathbf{I}_{N_r} + \frac{\rho}{n_t}\mathbf{\Lambda}\right)$$

$$= \prod_{i=1}^{N_r}\left(1 + \frac{\rho}{n_t}\lambda_i\right)$$

(2.81)

where λ_i is the ith eigenvalue of $\mathbf{HH}^\dagger$. Finally, substituting (2.81) into (2.55) yields,

$$C = E\left[\sum_{i=1}^{N_r}\left(1 + \frac{\rho}{n_t}\lambda_i\right)\right] \text{ bits/s/Hz}$$

(2.82)

which is subject to constant transmit power. Equation (2.82) shows that, thanks to the properties of the logarithm, the channel capacity of a MIMO wireless communication system is the sum of capacities of n_r virtual SISO channels defined by the spatial eigenmodes of the matrix product $\mathbf{HH}^\dagger$.

According to (2.82), the channel capacity C attains its maximum value when an equal signal-to-noise ratio of ρ/n_t is allocated to each virtual channel i.e., the n_r eigenmodes of the channel matrix $\mathbf{H}$ are all equally effective. By the same token, the capacity C is minimum when there is a single virtual channel (i.e., all the eigenmodes of the channel matrix $\mathbf{H}$ are zero except for one). The capacities of actual wireless links lie somewhere between these two extremes.

Specifically, as a result of fading correlation encountered in practice, it is possible for the existence of a large disparity among the eigenvalues of $\mathbf{HH}^\dagger$; that is, one or more of the eigenvalues $\lambda_1, \lambda_2, \ldots, \lambda_{nr}$ may assume small values. Such a disparity has a strong detrimental effect on the capacity of the wireless link compared to the maximal condition for which all the eigenvalues of $\mathbf{HH}^\dagger$ are equal. A similar effect may also arise in a Rician fading environment when the line-of-sight component is quite strong (i.e., the Rician factor is greater than 10 dB, say), in which case one eigenmode of the channel is dominant. For example, when the angle spread of the incoming radio waves impinging on a linear array of receive antennas is reduced from $60°$ to $0.6°$, the complementary cumulative distribution function of a MIMO wireless communication system with a (7,7) antenna configuration degenerates effectively to that of a (7,1) system.

The eigendecomposition of the log-det capacity formula given in (2.82) applies to a MIMO wireless link with $N_r < n_t$. For $n_t < N_r$, subject to constant transmit power, the capacity of the link may be expressed in the following two equivalent forms:

$$C = \log_2 \left\{ \det\left(\mathbf{I}_{n_t} + \frac{\rho}{n_t}\mathbf{H}^\dagger\mathbf{H}\right) \right\}$$

$$= \sum_{i=1}^{n_t} \log_2 \left(1 + \frac{\rho}{n_t}\lambda_i\right)$$

(2.83)

where the λ_i are the eigenvalues of the alternative n_t-by-n_t matrix product $\mathbf{H}^\dagger\mathbf{H}$.

2.6 SUMMARY AND DISCUSSION

In this chapter, we discussed the capacities of different forms of space diversity schemes. The main idea behind these schemes is that two or more propagation paths connecting the receiver to the transmitter are better than a single propagation path. In historical terms, the first form of space diversity used to mitigate the multipath fading problem was receive diversity, involving a single transmit antenna and multiple receive antennas. By far the most powerful form of space diversity involves the use of multiple antennas at both the transmit and receive ends of the wireless link. The resulting configuration is referred to as a multiple-input, multiple-output (MIMO) wireless communication system, which includes receive diversity and transmit diversity as special cases of space diversity. The novel feature of a MIMO system is that in a rich scattering environment, it can provide high spectral efficiency, which may be simply explained as follows. The signals transmitted simultaneously by the transmit antennas arrive at the input of each receive antenna in an uncorrelated manner due to the rich scattering mechanism of the channel. The net result is a spectacular increase in the spectral efficiency of the wireless link. Most importantly, the spectral efficiency increases roughly linearly with the number of transmit or receive antennas, whichever is smaller. This important result assumes that the receiver has knowledge of the channel state. The spectral efficiency of the MIMO system can be further enhanced by including a feedback channel from the transmitter to the receiver whereby the channel state is also made available to the transmitter and with it the transmitter is enabled to exercise control over the transmitted signal.

One last comment is in order. The discussion of channel capacity presented in the chapter focused on single-user MIMO (SU-MIMO) links. The derivation of MIMO channel capacity is much easier for single users than multiple users. Capacity formulas are known for many single-user MIMO cases, whereas the corresponding multiuser ones are unsolved. Simply put, very little is known about the channel capacity of multiuser MIMO links unless the channel state is known at both the transmitter and the receiver. Although the wireless systems in current use cater to the needs of multiple-user MIMO (MU-MIMO), the focus on single users may be justified in limited cases when MIMO systems are used with traditional time-division multiple access (TDMA) and frequency-division multiple access (FDMA). The SU-MIMO capacity results presented in this chapter will provide the capacity results.

3

BLAST ARCHITECTURES

Transmission techniques for MIMO wireless communications may be considered under two broadly defined categories:

1. Unconstrained signaling techniques, exemplified by the so-called BLAST architectures, whose aim is to increase the channel capacity by using layered space-time codes designed using standard channel codes.
2. Space-time codes, whose aim is the joint channel encoding of multiple transmit antennas to increase system diversity.

BLAST architectures are discussed in this chapter. BLAST is the first fundamental design of a multielement antenna system using building blocks of separately coded/decoded one-dimensional subsystems and interference-cancellation receivers. Despite the n_t-dimensional (n_t-D) transmitted signal, the layered space-time structures are designed to avoid the exponential growth of the processing complexity with the n_t-D transmitted signal, yet it can achieve a significant percentage of the matrix channel capacity. Theoretical development of BLAST concepts is based on the premise that error-free decisions are available in the interference-cancellation stage, which, in turn, implies the assumption that the system requires powerful and arbitrarily long channel codes and perfect decoding.

This chapter discusses four BLAST architectures:

- The diagonal layered space-time architecture, hence the term *diagonal-BLAST* or *D-BLAST* [43]. D-BLAST uses the concept of successive independent coded "diagonals" as shown in Figure 3.1(a). Importantly, these diagonals

Space-Time Layered Information Processing for Wireless Communications,
By Mathini Sellathurai and Simon Haykin
Copyright © 2009 John Wiley & Sons, Inc.

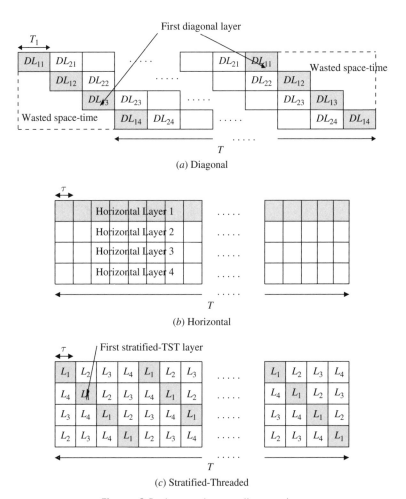

(a) Diagonal

(b) Horizontal

(c) Stratified-Threaded

Figure 3.1 Layered space-time codes.

offer a means of muting mutual interference while enabling one-dimensional (1-D) coding and decoding. In principle, the D-BLAST architecture attains the MIMO channel capacity. A major drawback of D-BLAST, however, is the requirement of short but efficient diagonal layer coding schemes to reduce its boundary space-time edge wastage. In practice, one cannot use a too-short diagonal, as that might not allow adequate time to support a powerful bandwidth-efficient code with forward error-correcting capabilities.

- The vertical-BLAST or V-BLAST, was the first practical system demonstrated. In V-BLAST, every antenna transmits its own independently encoded horizontal layer of data (see Figure 3.1b). Although V-BLAST eliminates the need for short block codes, the capacity achieved by V-BLAST for cases where $n_t \geq n_r$ is substantially lower than the Shannon limit [44].

- SD-BLAST is an alternative communication approach drawing on a more refined view of space-time suitable for BLAST systems with $n_t \geq n_r$. The proposed approach introduces a new feature called stratification, which we call stratified diagonal BLAST, or SD-BLAST. Moreover, in contrast to the successive diagonals in D-BLAST, in SD-BLAST the layering is done in a manner similar to the threaded space-time (TST) layering introduced in [40], but with an internal stratification feature. In effect, in SD-BLAST, a long diagonal has been replaced by a number of stratified layers that wind many times around the space-time. Figure 3.1(c) shows the TST layering where the internal stratification feature is not explicitly shown. The stratification feature is shown in Figure 3.10. At the receiver, the interference is muted with a successive decoding (so-called onion peeling) approach using a 1-D decoder stripping off stratum by stratum [127].

- Multirate layered space-time coded systems with successive decoding and interference cancellation (SDIC) receivers can be viewed as a class of diagonal layered space-time coded systems, with each of the layers being encoded independently and with different rates subject to equal per-layer outage probabilities. The layered space-time coding structure is the same one described in Figure 3.1(c) but the layers are coded using different rates. For sufficiently large numbers of transmit and receive antennas, the system can achieve near capacity in quasi-static fading environments [128].

3.1 BLAST ARCHITECTURE

As pointed out previously, the BLAST architecture consists of multiple antennas at both the transmitting and receiving ends of the system, as illustrated in Figure 3.2.

In this system, information-bearing signals are divided into multiple substreams and an array of antennas is used to launch the substreams simultaneously, using the same frequency bandwidth, with the total transmitted power always being held constant. At the receiving end, the transmitted signals are picked up by a receiving antenna array. Each antenna element of the array receives all the

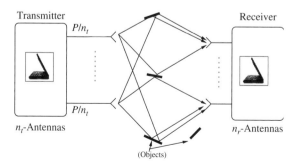

Figure 3.2 A BLAST system.

transmitted signals as one superimposed signal. Even though the signals are transmitted in the same frequency band, the signals from the different transmit antennas are located at different points in space, and each signal is scattered differently. Hopefully, the received signal at each receive antenna element still contains useful information about the transmitted signal. Since BLAST does not require additional spectrum resources to transmit parallel substreams (i.e, each antenna operates in a co-channel manner), the BLAST architecture is spectrally efficient. However, the spatial multiplexing and simultaneous use of the same portion of the spectrum lead to co-antenna interference, which is the major source of channel impairment in the BLAST architecture. Figure 3.3 illustrates the effect of co-antenna interference for one, two, and eight simultaneous transmissions with single reception. For the eight simultaneous transmissions, the "eye" of the received signal is closed, which illustrates the major limitation of multiple transmission.

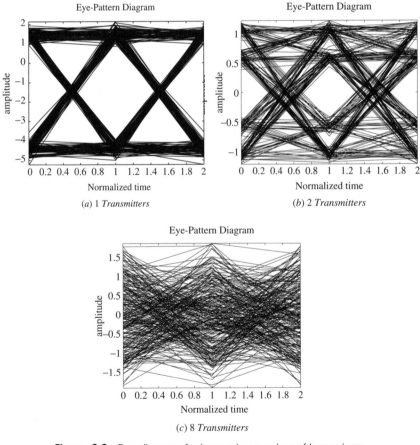

Figure 3.3 Eye diagram for increasing number of tranceivers.

Optimal codecs for MIMO schemes are multidimensional and can be found through exhaustive search methods, as in the maximum likelihood algorithms [65]. However, search complexity increases exponentially with the number of transmit antennas, the number of bits per modulation symbol, and the burst size. For multidimensional codes, the number of $N_s \times n_t$ matrices $\mathbf{X}$ needed in a codebook can be large. The following example illustrates the computation needed for the optimal codec solution for a (n_t, n_r) system.

The fundamental question raised by Foschini is: *Can one construct a BLAST system whose capacity scales linearly with the number of transmit antennas, using building blocks of n_t separately coded 1-D subsystems of equal capacity?* The motivation for raising this question is twofold:

- The space-time transmitter can be designed with the already developed 1-D codec technology.
- Low-complexity interference cancelation receivers can be used.

This question was answered in [43]: Using D-BLAST, we can indeed achieve the Shannon capacity. D-BLAST processes 2-D signals in MIMO systems using the already developed 1-D codec technology. The first dimension refers to time and the second dimension refers to space. Note that, in general, the 1-D code involves many dimensions over the time domain.

3.2 DIAGONAL BLAST

The innovative feature of the D-BLAST transmitter is the space-time encoding structure constructed with n_t diagonal layering 1-D coded subsystems of equal capacity, which permit decoding complexity to grow linearly with the number of transmit antennas. However, this architecture requires the use of diagonal layering. The space-time wasted at the start and end of a burst is significant for a practical burst length of few hundred symbols, even though this boundary waste becomes negligible as the burst length increases. Note that the use of a short packet size is important in wireless communications for two reasons:

1. Long packets require channel tracking inside a packet since the wireless channel varies with time.
2. Wireless communication is usually delay-limited.

3.2.1 The Diagonal-Layered Space-Time Codes

Figure 3.4 illustrates the D-BLAST transmitter. A data stream is demultiplexed into n_t data substreams of equal rate, and each data substream is encoded independently using block encoders. Rather than committing each of the n_t coded substreams to an antenna, the bit stream per antenna association is periodically cycled [43].

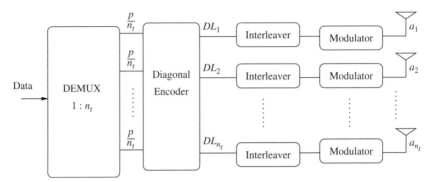

Figure 3.4 D-BLAST architecture.

The resulting diagonal layer codes due to the proposed periodic cycling of sub-streams are illustrated in Figure 3.1(a). These independent coded diagonal layers, referred to as diagonal-layer-space-time, $DL_i, i = 1, 2, \ldots, n_t$, arise in describing the space-time disposition of 1-D encoded message constituents that are placed diagonally across the antennas from which they radiate.[1] We denote the sublayers radiated from the jth antenna as DL_{ij}, forall $\forall_i$. It should be noted that the sublayers DL_{ij} may contain several coded symbols if $T > n_t$, where the number of vector symbols in a message is denoted by T. Note the edge wastage that is associated with this architecture.

The n_t encoded diagonally layered substreams denoted by $\{DL_i\}_{i=1}^{n_t}$ share

- a balanced presence over all n_t individual channel paths to the receiver; thus, each diagonal layer $\{DL_i\}_{i=1}^{n_t}$ has the same capacity $C_{DL_i} = C_D/n_t$, where C_D is the total capacity of D-BLAST, and
- a set of sublayers, DL_{ij}, with various SNRs; thus, the capacity obtained by each diagonal layer is equal to the sum of the $SNR_k s$ of n_t sublayers.

3.2.2 Serial Interference Cancellation Decoder

In the detection process, each diagonal layer constitutes a complete code word; thus, the diagonal layers are "peeled off" one by one from outside to inside. At any time, the previously peeled diagonals (the outer diagonals) have been removed as a source of interference in detecting bits in subsequent diagonals. Interference from the diagonal layers that will be peeled off later (the interfering inner diagonals) is suppressed by using minimum mean square error processing. Note that the capacity of each diagonal layer is identical, since each layer has a virtually identical structure after removal of the interference from its preceding outer diagonal layers. Therefore, it is sufficient to analyze the capacity of a single diagonal layer.

---

[1]The first diagonal layer is formed by the first and last diagonals, as illustrated in Figure 3.1(a).

At the receiving end, the decoupling process for each of the n_t layers involves a combination of nulling out the interference from yet undetected signals and canceling out the interference from already detected signals. This involves two steps:

1. Assuming that the receiver has detected the last signals $x_{i+1}, \ldots, x_{n_t}$ correctly; then we can cancel the interference from the decided components of **x**.

   To be specific, we write the received signal as follows:

$$\mathbf{y} = (x_1\mathbf{h}_1 + x_2\mathbf{h}_2 + \ldots + x_{i-1}\mathbf{h}_{i-1}) + x_i\mathbf{h}_i + (x_{i+1}\mathbf{h}_{i+1} + x_{i+2}\mathbf{h}_{i+2}$$
$$+ \ldots + x_{n_t}\mathbf{h}_{n_t}) + \mathbf{v} \tag{3.1}$$

   The last bracketed sum involves only correctly detected signals and is subtracted from the received signal to get the modified received signal

$$\mathbf{y}_i = (x_1\mathbf{h}_1 + x_2\mathbf{h}_2 + \ldots + x_{i-1}\mathbf{h}_{i-1}) + x_i\mathbf{h}_i + \mathbf{v} \tag{3.2}$$

2. Interference nulling using the whitened matched filters. The current desired signal is x_i, and the remaining signals $[x_1 + x_2 + \ldots + x_{i-1}]$ are interferences. The channel vector corresponding to the desired signal is $\mathbf{h}_i$.

   When the receiver processing weights for suppressing the remaining interferences in each layer are chosen to maximize the output SNR, instead of using the nulling process presented in [43], one can achieve the maximum SNR available, known as the matched filter bound:

   • Spatial-whitening process (whitening of interferences and noise) [6]:

$$\hat{x}_i = \mathbf{h}_i^\dagger \Psi_{i-1}^{-1/2} \mathbf{y}_i + v_i \tag{3.3}$$

   where the vector v_i has statistically independent complex Gaussian components with mean zero and variance one, and Ψ_{i-1} is the variance-covariance matrix of interference and additive white Gaussian noise, which is given by

$$\Psi_{i-1} = \sum_{k=1}^{i-1} \mathbf{h}_k\mathbf{h}_k^\dagger + \mathbf{I}_{n_r} \tag{3.4}$$

   • The signal-to-noise ratio of the ith signal is defined by

$$\Upsilon_i = \left[\mathbf{h}_i^\dagger \Psi_{i-1}^{-1/2} x_i\right]\left[\mathbf{h}_i^\dagger \Psi_{i-1}^{-1/2} x_i\right]^\dagger$$
$$= \mathbf{h}_i^\dagger \Psi_{i-1}^{-1} \mathbf{h}_i \frac{\rho}{n_t} \tag{3.5}$$

   where Υ_i is the maximum achievable SNR by any space-time processing receiver and is known as the matched filter bound, and ρ is the average SNR at each receiver input.

3.2.3 Capacity: Diagonal Layering of Space-Time

We analyze the ith diagonal layer composed of n_t- sublayers denoted by $\{DL_{ik}\}_{k=1}^{n_t}$. As described in Figure 3.1(a), the entries below and above the first diagonal layer are zeros. Subsequently, after removal of the interferences due to the outer diagonals, each sublayer of any diagonal layer will see a different level of the SNR. If we denote $\{\Upsilon_k\}_{k=1}^{n_t}$ as the generalized output SNR of sublayers, the D-BLAST random information rate is the summation of capacity of all the sublayers:

$$C_D(\mathbf{H}) = \sum_{k=1}^{n_t} \log_2[1 + \Upsilon_k] \qquad (3.6)$$

where

$$\Upsilon_k = \frac{P}{n_t \cdot \sigma^2} \left(\mathbf{h}_k^\dagger \left[\mathbf{I}_{n_r} + \frac{P}{n_t \cdot \sigma^2} \underline{\mathbf{H}}_k \underline{\mathbf{H}}_k^\dagger \right]^{-1} \mathbf{h}_k \right) \qquad (3.7)$$

where $\underline{\mathbf{H}}_k$ is derived from $\mathbf{H}$ by deleting its columns corresponding to indices $\{k, k+1 \ldots, n_t\}$; the dagger denotes Hermitian transposition. Details of the proof that the capacity in is (3.6) pertaining to D-BLAST as described here can be found in [44]. Reference [44] draws heavily on [6], which addresses a related but different problem. Thus, we refer to the performance of D-BLAST as the upper bound on capacity that BLAST architectures can reach using progressively more powerful channel codes. This capacity of D-BLAST architecture is in fact the Shannon capacity. A summary of the proof in [6] is provided in the appendix at the end of the chapter.

3.3 VERTICAL BLAST (V-BLAST)

To reduce the computational difficulty of D-BLAST, Wolniansky et al. [56] proposed a simplified version of BLAST known as vertical BLAST or V-BLAST, which is the first practical implementation of MIMO wireless communications in demonstrating a spectral efficiency as high as 40 bits/s/Hz in real time. In V-BLAST, the incoming binary data stream is first demultiplexed into n_t substreams, with each substream being encoded independently and mapped onto an antenna of its own for transmission over the channel. As far as the transmitter is concerned, the net result is the conversion of the incoming binary data stream into a vertical vector of encoded substreams, hence the name V-BLAST. The block-labeled vertical encoder in the transmitter part of the high-level block diagram of Figure 3.1(b) accounts for the demultiplexing and decoding operations. Comparing Figure 3.1(b) for V-BLAST with Figure 3.1(a) for D-BLAST, we see that in V-BLAST there is no cycling over time, hence the significant reduction in system complexity. Moreover, in the V-BLAST transmitter, every antenna transmits its own independently coded substream of data, as illustrated in Figure 3.1(b). Although, in so doing, V-BLAST eliminates the space-time edge wastage plaguing

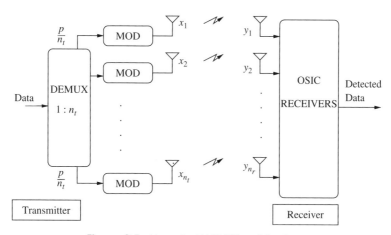

Figure 3.5   Uncoded V-BLAST architecture.

D-BLAST, the outage capacity achieved by V-BLAST for antenna configurations with $n_t \geq n_r$ is substantially lower than that of D-BLAST.

Turning next to the receiver, the function of the block labeled OSIC in Figure 3.5 is to detect the original data stream. A straightforward method to perform the detection is to use an adaptive antenna array. In this approach, each substream is considered to be a desired signal and the remaining substreams are considered to be interferers. The weights of the antenna array are adjusted in an iterative manner so as to place nulls along the interferers and thereby minimize a cost function defined as the mean square value of the error signal (i.e., the difference between the desired response and the actual response of the antenna array). In V-BLAST, superior performance is achieved by using a nonlinear technique called the ordered serial interference cancellation (OSIC) detector, which exploits the timing synchronism between the receiver and transmitter. This OSIC scheme is based primarily on successive interference cancellation using hard decisions and a new scheme to find the order in which the substreams are to be detected for optimum performance. In this detection scheme, the substreams are first sorted according to the strength of each subchannel. Then the strongest substream is detected and extracted from the total received signal, and the sorting and extraction are repeated until all the substreams are detected. In summary, the OSIC detection scheme involves the following sequence of operations:

1. Order determination, in which the n_t received substreams are to be detected, in accordance with the postdetection SNRs of the individual substreams.
2. Detection of the substream, starting with the largest SNR.
3. Signal cancellation, where the effect of the detected substream is removed from subsequent substreams.
4. Repetition of steps 1 through 3 until all the n_t received substreams have been detected.

The procedure is nonlinear due to two factors. First, the estimated signal cancellation process is itself nonlinear. Second, the detection step involves a quantization (slicing) operation that is appropriate to the signal constellation used in the transmitter. Another noteworthy point is that the nulling and cancellation step is performed by exploiting the channel matrix **H**, which is estimated at the receiver through supervised training aided by sending a training sequence from each transmit antenna. Specifically, the nulling process suppresses the interference. This process is followed by cancellation of the estimated signal in question. The relatively simple design of the receiver, as summarized herein, makes V-BLAST a candidate for the next generation of wireless communications.

3.3.1 OSIC Detection Algorithm (56)

We consider the MIMO received signal model given in (2.9). Let the detection order

$$s = \{k_1, k_2, \ldots, k_{n_t}\} \tag{3.8}$$

be a permutation of integers $1, 2, \ldots n_t$. The detection process has three major steps:

1. From the soft decision statistics for the k_i^{th} signal as follows:

$$y_{k_i} = \mathbf{w}_{k_i}^\dagger \mathbf{y}_i \tag{3.9}$$

where the weight vector $\mathbf{w}_{k_i}$ is obtained by using well-known linear filters such as zero-forcing (ZF) or minimum mean-squared-error (MMSE) whitened matched filtering, [67].

Problem 1 (ZF): *Given the model (2.9), estimate the user information sequence* **x** *by minimizing cost function*

$$\min_{\mathbf{x}} \|\mathbf{y} - \mathbf{H}\mathbf{x}\|_{\mathbf{R}_v}^2 \tag{3.10}$$

where $\mathbf{R}_v = E\{\mathbf{v}\mathbf{v}^\dagger\} = \sigma^2 \mathbf{I}.$

Solution to Problem 1 (ZF): *The solution to this problem is given by*

$$\hat{\mathbf{x}} = \mathbf{H}^+ \mathbf{y} \tag{3.11}$$

The corresponding weight matrix is

$$\mathbf{w} = \mathbf{H}^+ \tag{3.12}$$

where $\mathbf{H}^+ = (\mathbf{H}^\dagger \mathbf{H})^{-1} \mathbf{H}^\dagger$ *is the pseudoinverse of matrix* **H** *[57]. Note that the (ZF) weight vector computation is inherently ill-posed.*

Problem 2 (MMSE): *Given model (2.9), estimate the user information sequence* **x** *by minimizing the following cost function:*

$$\min_{\mathbf{x}} \|\mathbf{x} - \hat{\mathbf{x}}\|^2 \tag{3.13}$$

Solution to Problem 2 (MMSE): *The solution to this problem is given by*

$$\hat{\mathbf{x}} = (\mathbf{H}^\dagger \mathbf{H} + \sigma^2 \mathbf{I})^{-1} \mathbf{H}^\dagger \mathbf{y} \tag{3.14}$$

The corresponding weight matrix is

$$\mathbf{w} = (\mathbf{H}^\dagger \mathbf{H} + \sigma^2 \mathbf{I})^{-1} \mathbf{H}^\dagger \tag{3.15}$$

2. Obtain the hard decisions $\hat{x}_{k_i}$ by finding the nearest point to y_{k_i} in the constellation coordinates which minimizes the Euclidean distance based on the soft decisions

$$\hat{x}_{k_i} = Q(y_{k_i}) \tag{3.16}$$

3. Cancel the signal component due to $\hat{x}_{k_i}$ from the received vector $\mathbf{y}_i$, obtaining

$$\mathbf{y}_{i+1} = \mathbf{y}_i - \hat{x}_{k_i} \mathbf{h}_{k_i} \tag{3.17}$$

where $\mathbf{h}_{k_i}$ denotes the k_ith column of **H**.

Ordering of Substreams using the Postdetection SNR The postdetection SNR for the k_ith detected component of **x** is given by

$$\rho_{k_i} = \frac{E\{|x_{k_i}|^2\}}{\sigma^2 \|\mathbf{w}_{k_i}\|^2} \tag{3.18}$$

where the expectation operator E is taken over the constellation points. The larger the ρ_{k_i}, the smaller the nulling weights $\mathbf{w}_{k_i}$ will be. Thus, the postdetection SNR is maximized by the smallest nulling weight; detecting and canceling the signal with the smallest nulling weight improves the algorithm. V-BLAST with optimal ordering based on the smallest weight is described below [56].

- Initialization:

$$i \rightarrow 1$$
$$\mathbf{G}_1 = \mathbf{H}^+ \tag{3.19}$$

- Recursion:

$$k_i = \arg\{ \min_{j \notin (k_1,\dots,k_{i-1})} \|(\mathbf{G}_i)_j\|^2 \}$$
$$\mathbf{w}_{k_i} = ((\mathbf{G}_i)_{k_i})^T$$

$$y_{k_i} = \mathbf{w}_{k_i}^{\dagger} \mathbf{y}_i$$

$$\hat{x}_{k_i} = Q(y_{k_i}) \qquad\qquad (3.20)$$

$$\mathbf{y}_{i+1} = \mathbf{y}_i - \hat{x}_{k_i} \mathbf{h}_{k_i}$$

$$\mathbf{G}_{i+1} = \mathbf{H}_{\overline{k_i}}^{+}$$

$$i \leftarrow i + 1$$

where $(\mathbf{G}_i)_{k_i}$ is the k_ith row of $\mathbf{G}_i$, $\mathbf{H}_{\overline{k_i}}$ denotes the matrix obtained by deleting columns $k_1, k_2, \ldots, k_i$ of $\mathbf{H}$, and $(\cdot)^{+}$ denotes the pseudoinverse [57].

Remark 1 *To perform the optimal ordering process used in V-BLAST, the pseudo-inverse $\mathbf{G}_i$ must be computed for $i = 1, 2, \ldots, n_t$, which is sensitive with respect to channel matrix $\mathbf{H}$ that has correlated columns: a small change in $\mathbf{H}$ may cause large and unpredictable variations in $\mathbf{G}_i$. Note that $\mathbf{G}_i$ is computed by deleting $n_t - i + 1$ columns of $\mathbf{H}$. This sensitivity of the pseudoinverse grows linearly with $n_t - Rank(\mathbf{H})$ [57, 31]. Note also that this sensitivity may be reduced by setting the smaller singular values to zero, but determining the numerical rank of this problem is challenging. Therefore, the ordering process may fail when $n_t \geq n_r$. Typically, for $n_t \leq n_r$, the columns of $\mathbf{H}$ may be assumed to be independent.*

3.3.2 Improved V-BLAST

Many modifications have been proposed to improve V-BLAST. A channel-based adaptive group detection (AGD) combined with OSIC is proposed in [31]. In the OSIC-AGD algorithm, the ZF/MMSE-based interference nulling/suppression is replaced with the AGD algorithm.

The AGD algorithm contains three steps: grouping based on the channel information, subspace projection, and maximum likelihood (ML) search within each group:

1. Several layers are grouped for processing together.
2. ML detection is performed within the groups.
3. Interference is canceled in such a way that only part of the search results from each group is used as decisions.

Repeat steps 1 and 2 for the unprocessed layers until all the layers have been processed. The complexity of the ML detection over all n_t transmitters requires $N_I^{n_t}$ searches, where N_I is the size of the constellation used, which is beyond the limit of most systems today. Group detection is carried out within a group of n_g, where $n_g \ll n_t$; therefore, the search complexity is proportional to $N_I^{n_g}$ and can be kept low by keeping the group size n_g small.

3.3.3 Coded V-BLAST

Three coded V-BLAST schemes exist in the literature:

1. V-BLAST with horizontal encoder V-BLAST-I
2. V-BLAST with vertical encoder V-BLAST-II
3. V-BLAST with concatenated coder V-BLAST-III

In the horizontal encoding scheme, the substreams are independently encoded using 1-D channel codes, as shown in Figure 3.6. The incoming data are first divided into $\mathbf{b}_1, \mathbf{b}_2, \ldots, \mathbf{b}_{n_t}$ and each part is encoded separately, then interleaved and symbol-mapped to generate the parallel substreams. The receiver performs OSIC detection and decoding. In this scheme, the receiver cancels interference using decisions of previously decoded signals.

One possible shortcoming of horizontal coding compared to diagonal coding is that overall performance may be dominated by the weakest layer, particularly the first decoded layer, because it has the lowest diversity in V-BLAST decoding. Note that the first detection layer has the minimum SNR among the n_t possible layers, since it experiences $n_t - 1$ interferences; the second detection layer has only $n_t - 2$ interferences; and so on. The ordering process used in the V-BLAST OSIC receiver maximizes the minimum SNR by maximizing the post-detection SNR [43].

In the vertical encoding approach (see Figure 3.7), a single code is used to encode all the signals, and the coded information bits are demultiplexed across the n_t parallel streams $\mathbf{x}_1, \mathbf{x}_2, \ldots, \mathbf{x}_{n_t}$. At the other end, the receiver first decouples the n_t data streams through interference nulling/cancellation, as described for uncoded V-BLAST, then multiplexes and decodes all the n_t substreams as one information block. In this scheme, the effective SNR averaging across the antenna array may be achieved because all the layers are coded together. A serious drawback of vertical coding is that only the undecoded decisions are provided to the interference canceler since decoding cannot be done until all the layers are processed. Thus, vertical

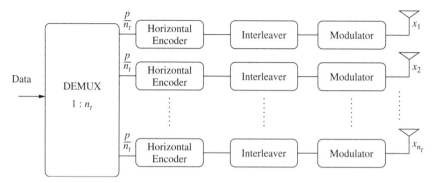

Figure 3.6 Horizontally coded V-BLAST architecture.

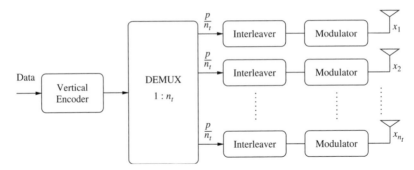

Figure 3.7 Vertically coded V-BLAST architecture.

coding is more prone to decision errors than horizontal coding. Consequently, the rate of vertical encoding approaches $R_h < C_V$, where C_V is defined in (3.24)

The third coding is the concatenated approach shown in Figure 3.8. The horizontal coding layers are designed, typically with convolutional codes as inner codes, to facilitate OSIC detection using decoded decisions, whereas the vertical coding, with Reed-Soloman outer code, is included for SNR averaging across the antennas. A drawback of this scheme is a reduced information rate due to the concatenation of two coding schemes.

In general, to minimize the effect of decision errors and to improve the joint detection and decoding gain, it is assumed that turbo processing is incorporated as second-round processing, as illustrated in Figure 3.9. Iterative detection used in second-stage processing could improve the performance if the performance due to the first processing is far from the capacity C_V.

3.3.4 Limitations of V-BLAST

1. The V-BLAST detection algorithm has a computational complexity $O(N^4)$, where N is the number of transmit and receive antennas. This is a major

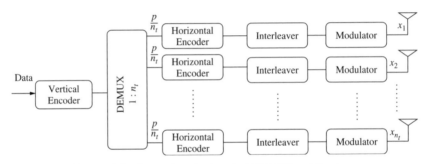

Figure 3.8 Concatenated coded V-BLAST architecture.

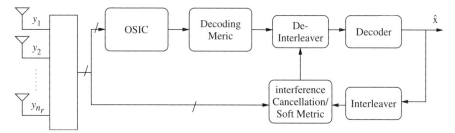

Figure 3.9 Second-stage iterative receiver for V-BLAST.

issue to be considered. The V-BLAST scheme uses a sequential nulling and interference cancellation strategy: decoding the strongest signal first, then canceling the interference due to the decoded signal; decoding the strongest signal of the remaining signals; and so on. In this receiving scheme, the optimal order of detection and the nulling vector have to be determined at each decoding step, which can be computationally expensive and numerically unstable. We may estimate the optimal ordering using the singular value decomposition (SVD), which is the most stable way of computing the pseudoinverse. For an $n_r \times n_t$ channel matrix, whose entries are complex values, we note the following ([57], [65]):

- The singular value decomposition has a complexity of $2n_r{}^2 n_t + 11n_t{}^3$.
- The pseudoinverse has a complexity of $2n_r n_t^2$.
  Since these steps have to be repeated n_t times in the sequential detection procedure with the matrix sizes of dimension $(n_r) \times k, k = n_t, n_t - 1, \ldots, 1$, the total complexity is

$$\sum_{k=1}^{n_t} \{2(n_r + k)^2 k + 11k^3 + 2(n_r + k)k^2\} = n_r{}^2 n_t{}^2 + 2n_r n_t^3 + \frac{15}{4} n_t{}^4$$

(3.21)

Therefore, for an (N,N)-BLAST, the computations are on the order of N^4.

2. V-BLAST demands more receive antennas than transmit antennas. In contrast, the ability to work with fewer receivers than transmitters is necessary in most cellular communications systems since the base station is typically designed to handle more antennas than the mobile unit.

3. V-BLAST does not use any transmit diversity inherent in the MIMO system. It has no built-in space-time codes to overcome deep fades from any of the transmit antennas and suffers from the problem of error propagation: a wrong decision made at the output leads to a higher probability of error on subsequent decisions.

4. D-BLAST and V-BLAST are quite sensitive to channel estimation errors since they use hard-decision-based interference cancellation receivers. Note that, in practice, the channel is learned by the receiver by using short training

sequences transmitted by the transmitter with each packet of information symbols.

Remark 2 Complexity Reduction: *A cost-efficient and numerically stable "square-root" algorithm has been proposed for the V-BLAST architecture. This square root algorithm has the following favorable features:*

- *The computational cost is reduced to $0.7n$, where n is number of transmitters.*
- *It uses only orthogonal transformations and is numerically stable.*
- *It can be implemented using fixed-point digital signal processing (DSP) hardware, whereas the original V-BLAST algorithm requires floating-point DSP.*

3.3.5 Capacity: Vertical Layering of Space-Time

In this architecture, the primitive bit stream is multiplexed into n_t horizontal layers, as shown in Figure 3.1(*b*). Each horizontal layer is independently encoded using 1-D channel codes and transmitted using a different antenna.

The 1-D receiver processing proposed for V-BLAST is OSIC. In the OSIC receiver, the layers will be decoded in a permuted order $\{k_1, k_2, \ldots, k_{n_t}\}$ that maximize the rate of the weakest decoding layer in the decoding process. Note that there are $n_t!$ possible permutations to be tested to find the optimal ordering.

Assume that the best decoding order that maximizes the capacity of the worst decoding layer is $\{k_1, k_2, \ldots, k_{n_t}\}$; the instantaneous capacity of V-BLAST is given as the random variable [43]

$$C_{Vn_t}(\mathbf{H}) = n_t \times \min_{m \in [1,2,\ldots,n_t]} \log_2[1 + \Upsilon_m] \tag{3.22}$$

where C_{Vn_t} is the capacity of V-BLAST when using all n_t transmit antennas, and the SNR denoted by Υ_m is given by

$$\Upsilon_m = \frac{P}{n_t \cdot \sigma^2} \left(\mathbf{h}_{k_m}^\dagger \left[\mathbf{I}_{n_r} + \frac{P}{n_t \cdot \sigma^2} \overline{\mathbf{H}}_{k_m} \overline{\mathbf{H}}_{k_m}^\dagger \right]^{-1} \mathbf{h}_{k_m} \right) \tag{3.23}$$

where $\overline{\mathbf{H}}_{k_m}$ is derived by deleting columns corresponding to indices $\{k_1, \ldots, k_{m-1}\}$ of channel matrix $\mathbf{H}$.

It must be stressed that with V-BLAST a subset of the transmitter antennas can give a better performance than the use of all the available n_t. Therefore, the capacity can be further optimized with respect to the number of transmit antennas that should be used for actual transmission

$$C_V(H) = \max_{m \in [1,2,\ldots,n_t]} m \times \max_{i \in [1,2,\ldots,\frac{!n_t}{!m \cdot !(n_t - m)}]} C_{Vm_i} \tag{3.24}$$

where we have evaluated the capacity for all possible combinations when using m antennas out of n_t transmit antennas to optimize the capacity of V-BLAST.

The corresponding outage capacity $C_V(\epsilon)$ is defined as

$$\epsilon = Pr\{C_V(H) < C_V(\epsilon)\} \qquad (3.25)$$

Note that at high SNRs, to find the optimum capacity of V-BLAST, it is sufficient to evaluate the $m^2/2$ myopic optimization levels introduced in [46].

3.4 STRATIFIED DIAGONAL BLAST^2 (SD-BLAST)

The key feature of D-BLAST is the diagonal layering with edge wastage. However, due to the edge wastage, a significant fraction of the information transmission is avoided. To compensate for this wastage, we will now consider a space-time architecture composed of n_t continuing diagonals, as shown in Figure 3.1(c), in which the space-time waste occurring in D-BLAST is eliminated. However, due to the simultaneous interferences in estimating (using 1-D signal processing techniques) each of the M continuing diagonals, the capacity that could be supported by this space-time architecture is limited. Thus, in SD-BLAST, besides the continuing diagonals, we will introduce a new feature called stratification. Stratification will effectively reduce the interferences by estimating the continuing diagonals in SD-BLAST. Later, we will show, both by theory and by Monte Carlo simulations, that the capacity performance of the SD-BLAST will improve with increasing stratification.

3.4.1 Transmitter

Figure 3.10 illustrates the SD-BLAST transmitter. First, a primitive bitstream is demultiplexed into n_t separate bitstreams of equal rate. Then each of these streams is further demultiplexed into L independent bitstreams of different rates (which we call L strata). It will be useful to introduce the term ply. A ply consists of n_t strata, each of which is transmitted by different antennas. We say that we have $L + 1$ plies because there are L signal plies and one AWGN ply. Subsequently, a message will be expressed as $n_t \cdot L$ 1-D coded blocks of data (strata). Alternatively, the coding can be done as L 1-D coded blocks (plies).

The stratified layers are cycled periodically over the n_t transmit antennas similar to the TST layering structure. In accordance with this cycling, we take the integer labels for transmit antennas to be the integers mod (n_t) and assume that the number of vector symbols in a message, T, is a multiple of n_t. In effect, a long diagonal has been replaced by a stratified layer that winds T/n_t times through the space-time architecture. The n_t congruent stratified layers are offset from each other in time by the time needed to send one symbol.

The $n_t \times T$ space-time code matrix is constructed by M codewords of length T, denoted by x_m, $m = 1, 2, \ldots, n_t$, as shown by the matrix

---

[2]We acknowledge the important contributions of Dr. G. J. Foschini in this section.

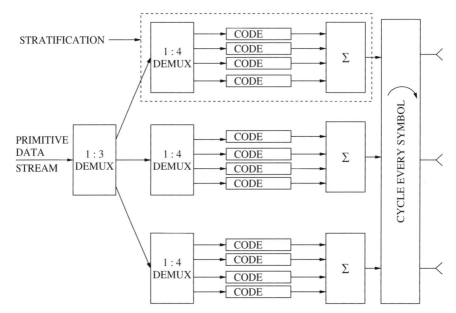

Figure 3.10 SD-BLAST transmitter.

$$\begin{pmatrix}
x_1(1) & x_{n_t}(2) & x_{n_t-1}(3) & \cdots & x_3(n_t-1) & x_2(n_t) & x_1(n_t+1) & \cdots & \cdots & x_2(T) \\
x_2(1) & x_1(2) & x_{n_t}(3) & \cdots & x_4(n_t-1) & x_3(n_t) & x_2(n_t+1) & \cdots & \cdots & x_3(T) \\
x_3(1) & x_2(2) & x_1(3) & \cdots & x_5(n_t-1) & x_4(n_t) & x_3(n_t+1) & \cdots & \cdots & x_4(T) \\
\vdots & \vdots & \vdots & \ddots & \vdots & \vdots & \vdots & \ddots & \ddots & \vdots \\
x_{n_t}(1) & x_{n_t-1}(2) & x_{n_t-2}(3) & \cdots & x_2(n_t-1) & x_1(n_t) & x_{n_t}(n_t+1) & \cdots & \cdots & x_1(T)
\end{pmatrix}$$

$$(3.26)$$

Note that each of the codewords is formed by superposition of L independent codewords of different rates; that is, $s_m(t) = \sum_{l=1}^{L} x_m^l(t)$, where we denote the mth stratum of the lth ply at time t by $x_m^l(t)$. Correspondingly, the lth ply at time t is denoted by $x^l(t)$. The received signal at time t is composed of L plies (equivalently, we can say $n_t \cdot L$ strata), given by

$$\mathbf{y}(t) = \mathbf{H}\sum_{l=1}^{L}\mathbf{x}^l(t) + \mathbf{v}(t) = \mathbf{H}\sum_{l=1}^{L}\begin{pmatrix} x_1^l(t) \\ x_2^l(t) \\ \vdots \\ x_{n_t}^l(t) \end{pmatrix} + \mathbf{v}(t) \qquad (3.27)$$

3.4.2 Receiver

At the receive array, the encoded bits are detected as follows. Each ply is to be regarded as a ring of an onion slice, as shown in Figure 3.11. In the figure, we assume that the first $L - 4$ plies are detected and canceled. Now the outer $(L - 3)$th ply must be detected first. Detection is followed by removal of the signals in the detected ply as a source of interference. In the process of detecting the bits in the strata of the exposed ply, the signal can be, but need not be, processed simultaneously and separately. After all the bits in the n_t strata that make up the outside ply are detected, assuming error-free detection, the interferences corresponding to these strata are subtracted from the received signal vector. The $(L - 2)$th ply is then exposed. The bits in the $(L - 2)$th ply are then detected, and detected strata signals are canceled, and so on, until finally the innermost $(L - 1)$th AWGN ply is left.

In preparing an exposed ply for detection, the n_r-D received vector, $y(t)$, which was received at time t, has already been processed, assuming perfect removal of the relatively outer ply. So, each inside ply is assumed to be free of interference from all outside plies. It may be helpful to think of a sequence of processed received vectors bootstrapped with $\mathbf{y}(t)$. So, with each major step in the processing sequence, the original received vector process $\mathbf{y}(t)$ sheds another ply's interference:

$$\mathbf{y}(t) \rightarrow \mathbf{y}^1(t) \rightarrow \mathbf{y}^2(t) \rightarrow \cdots \rightarrow \mathbf{y}^{L-1}(t) \qquad (3.28)$$

where all of the L vectors are an n_r-D function of time defined over the message duration: $\mathbf{y}^{i-1}(t)$ corresponds to the modified received signal used to detect the ith ply of the onion ring. Each vector is then reduced to a scalar using a weight vector, defined (up to an arbitrary complex scalar multiple) to maximize the signal-to-noise plus interference ratio (SINR) associated with this decision statistic. Maximization of SINR is a well-known process; see, e.g., [44].

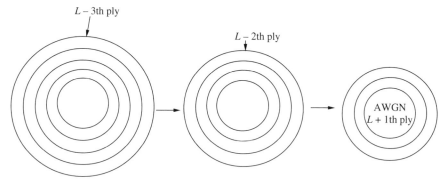

Figure 3.11 Peeling away of successive strata from the outside in. Here the last five layers ($L - 3$, $L - 2$, $L - 1$, L, and $L + 1$) are depicted as onion rings.

In SD-BLAST, as we mentioned earlier, each ply has to be coded with a rate seen at its level of mutual information. Therefore, we need to estimate the outage capacity of each stratum.

Remark 3 *According to the proposed detection scheme, the first (outermost) ply is decoded by treating the interference from the (L-1) inner plies as noise. The decoded transmitted codeword of the first ply is then subtracted from the received signal, and the second ply is decoded by treating the interference from the remaining plies as noise. Therefore, for a fixed channel matrix $\mathbf{H}$, the rate seen by the first ply subject to the proposed successive decoding method can be stated as the mutual information (MI) between the received signal and the first ply, which we denote as $I(\mathbf{y}; \mathbf{x}^1)$. Assuming that the first ply is detected error free, the rate seen by the second ply subject to the proposed successive decoding method is the MI between the received signal and the second ply given the codeword of the first ply (denoted by $I(\mathbf{y}; \mathbf{x}^2|\mathbf{x}^1)$), and so on. We have the following relationship between the MIs of each ply:*

$$
\begin{aligned}
I(\mathbf{y}; \mathbf{x}^1) &< I(\mathbf{y}; \mathbf{x}^2|\mathbf{x}^1) < \dots < I(\mathbf{y}; \mathbf{x}^L|\mathbf{x}^1, \mathbf{x}^2, \dots, \mathbf{x}^{L-1}) \\
I(\mathbf{y}; \mathbf{x}^1) &< I(\mathbf{y}^1; \mathbf{x}^2) < \dots < I(\mathbf{y}^{L-1}; \mathbf{x}^L)
\end{aligned}
\tag{3.29}
$$

The above inequalities are true for the following reasons:

1. *Each ply is cycled equally through all n_t transmit antennas, which averages the signal strength fluctuations in the spatial channels.*

2. *A ply with an increased number experiences decreased interference. In the following, we derive the corresponding formulas for MI subject to the maximization of SINR based on the receiver processing.*

3.4.3 Differential Rates of Individual Strata

We write the mathematical formula for the differential capacity added by each stratum subject to the proposed successive decoding or the so-called onion-peeling receiver processing. The formulas we derive will be useful in determining what can be achieved when the transmitter is informed only of the channel statistics.

First, we derive the formula for the bit rate that each stratum is capable of supporting when it is supplied with power $P/(n_t L \sigma^2)$ and subject to the proposed onion-peeling receiver processing, assuming that all the previous strata are decoded error-free so that error propagation does not occur. We investigate the received signal constituent, x_m^i, on the mth stratum of the ith ply at time t and determine how it is impaired. We do so for the case when the mth stratum exposed on the ith ply is to be detected following the detection and removal of interference from strata $1, 2, \dots, m-1$. Along with AWGN, this constituent, x_m^i, is impaired by all of the simultaneously transmitted strata with equal or higher indices that are not yet detected. The constituents of these interfering strata are $x_{m'}^l, i \le l \le L, 1 \le m' \le$

n_t, $m' \neq m$ when $l = i$. We express the received scalar signal x_m^i, along with its additive impairment, as

$$
\mathbf{y}^{m-1} = \left(\mathbf{h}_1, \mathbf{h}_2, \ldots, \mathbf{h}_{n_t}\right) \left[\begin{pmatrix} 0 \\ \vdots \\ 0 \\ x_m^i \\ 0 \\ \vdots \\ 0 \end{pmatrix} + \sum_{l=i}^{L,(mi)} \begin{pmatrix} x_1^l \\ x_2^l \\ \vdots \\ x_{n_t}^l \end{pmatrix} \right] + \mathbf{v} \qquad (3.30)
$$

The superscript (mi) on the summation on the right-hand side of (3.30) means that the ith stratum of the mth component is omitted from the sum because, in (3.30), the scalar x_m^i is regarded as the signal, not the interference. Recall that we are working with the normalized form of (2.9), so that the n_r-D noise vector $\mathbf{v}$ has complex i.i.d. normal (0,1) components.

In (3.30), the impairment is the n_r-D vector, ξ_{mi}, defined by

$$
\xi_{mi} = \sum_{j=1}^{n_t} \mathbf{h}_j \sum_{l=1}^{L,(mi)} x_j^l + v \qquad (3.31)
$$

For each value of $m = 1, 2, \ldots, n_t$, the variance-covariance matrix of this impairment of x_m^i is denoted by $\Sigma_{\xi_{mi}}$ (see (3.33)). Then we spatially whiten this vector impairment by multiplying the expression in (3.30) by the positive definite matrix $\Sigma_{\xi_{mi}}^{-1/2}$. The spatially whitened ith stratum of the mth component is given by $\Sigma_{\xi_{mi}}^{-1/2} \mathbf{h}_m x_m^i + \mu^m$, $m = 1, 2, \ldots, n_t$, where each of the vectors in the sequence of n_r-D vectors, $\{\mu^m\}_{m=1}^{n_t}$, has complex normal (0,1) components that are statistically independent of each other.

Next, employing E for expectation, we compute

$$
\Sigma_{\xi_{mi}} = E\left\{ \sum_{j=1}^{n_t} \mathbf{h}_j \left(\sum_{l=i}^{L,(mi)} x_j^l \right) + \mathbf{v} \right\} \left[\sum_{j=1}^{n_t} \mathbf{h}_j \left(\sum_{l=i}^{L,(mi)} x_j^l \right) + \mathbf{v} \right]^\dagger \qquad (3.32)
$$

$$
= \mathbf{I}_{n_r} + \left(\sum_{j=1}^{n_t} \mathbf{h}_j \mathbf{h}_j^\dagger \left(\frac{L-i+1}{L} \frac{P}{\sigma^2 \cdot n_r} \right) \right) - \mathbf{h}_m \mathbf{h}_m^\dagger \left(\frac{P}{\sigma^2 \cdot n_t} \right) \frac{1}{L} \qquad (3.33)
$$

where the matrix product $\mathbf{H}\mathbf{H}^\dagger = \sum_{j=1}^{n_t} \mathbf{h}_j \mathbf{h}_j^\dagger$ and $\mathbf{I}_{n_r}$ denotes the $n_r \times n_r$ identity matrix. Later we will need the large-L asymptotic form of (3.33). Consequently, for later use, we introduce $z = i/L$ and write

$$\Sigma_{\xi_{mi}} = \mathbf{I}_{n_r} + \left(\frac{P}{\sigma^2 n_t}\right) \mathbf{H}\mathbf{H}^\dagger (1 - z) + \underbrace{\frac{P}{\sigma^2 \cdot n_t}(\mathbf{H}\mathbf{H}^\dagger - \mathbf{h}_m \mathbf{h}_m^\dagger)\frac{1}{L}}_{\omega_L} \qquad (3.34)$$

where ω_L denotes the terms in the order of $1/L$. Therefore, the SINR of the ith stratum of the mth component is explicitly defined by

$$\Upsilon_{mi} = \mathbf{h}_m^\dagger \left(\mathbf{I}_{n_r} + \left(\frac{P}{\sigma^2 n_t}\right)\mathbf{H}\mathbf{H}^\dagger (1 - z) + \omega_L\right)^{-1} \mathbf{h}_m \frac{P}{\sigma^2 L n_t} \qquad (3.35)$$

As time progresses, the ith stratum in any diagonal layer experiences an SINR that is periodic with n_t. The capacity added by the ith stratum in any one diagonal layer, say the dth ($1 \le d \le n_t$), is then obtained by averaging the capacity contributions from the n_t transmit antennas. The differential capacity, C_i^d, added by the ith stratum when it is supplied with power $P/(L\sigma^2)$ is given by

$$C_i^d(\mathbf{H}) = \frac{1}{n_t}\sum_{m=1}^{n_t} \log_2[1 + \Upsilon_{mi}] \qquad (3.36)$$

This is a key equation that will be used for the computations of outage rates in Section 3.5.

3.4.4 Asymptotic Capacity of SD-BLAST as $L \to \infty$

By knowing the strata rates and using the communication means that we have described in the limit of large L, the log-det capacity is attained. To prove this, we derive the large-L asymptotic form of the strata capacities, still under the assumption that the transmitter is made privy to these capacities. First, we rewrite (3.36) as follows:

$$C_i^d(\mathbf{H}) = \frac{1}{n_t}\sum_{m=1}^{n_t} \log_2\left[1 + \mathbf{h}_m^\dagger \left(\mathbf{I}_{n_r} + \frac{P}{n_t\sigma^2}\mathbf{H}\mathbf{H}^\dagger (1 - z) + \omega_L\right)^{-1} \mathbf{h}_m \frac{P}{L n_t \sigma^2}\right] \qquad (3.37)$$

For large L, we employ the small-ϵ approximation for $\log_2(1 + \epsilon) \approx \epsilon / \ln 2$ so that the differential capacity contribution of the ith stratum is expressed as

$$C_i^d(\mathbf{H}) = \frac{P}{\sigma^2 \cdot L \cdot n_t^2 \cdot \ln 2}\sum_{m=1}^{n_t} \mathbf{h}_m^\dagger \left(\mathbf{I}_{n_r} + \frac{P}{n_t\sigma^2}\mathbf{H}\mathbf{H}^\dagger (1 - z)\right)^{-1} \mathbf{h}_m \qquad (3.38)$$

Note that the term $\omega_L \to 0$ as $L \to \infty$. Upon summing these incremental capacities C_i^d, from i equal to 1 to L, and then taking the limit $L \to \infty$, the summation becomes an integral over $[0, 1]$. For an integral of a function of the form $f(1 - z)$ over $[0, 1]$, if we substitute $\zeta = 1 - z$, it follows that $\int_0^1 f(1 - z)dz = \int_0^1 f(\zeta)d\zeta$.

So, in the large-L limit, we proceed by replacing ζ with $1 - z$ and $d\zeta$ associated with $1/L$. Therefore, the sum capacity, $\sum_{i=1}^{L} C_i^d(\mathbf{H})$, assumes the form

$$\lim_{L \to \infty} \sum_{i=1}^{L} C_i^d(\mathbf{H}) = \lim_{L \to \infty} \frac{P}{\sigma^2 \cdot n_t^2 \cdot \ln 2} \sum_{m=1}^{n_t} \int_0^1 \mathbf{h}_m^\dagger \left(\mathbf{I}_{n_r} + \frac{P}{n_t \sigma^2} \mathbf{HH}^\dagger \zeta \right)^{-1} \mathbf{h}_m d\zeta$$

(3.39)

Now the detection of n_t such diagonal layers is executed in parallel so that the total sum-capacity C_{SD}^L is obtained by multiplying the right-hand side of (3.39) by n_t. After the use of some minor linear algebra, the total capacity is found to be

$$\lim_{L \to \infty} C_{SD}^L(\mathbf{H}) = \frac{P}{\sigma^2 \cdot n_t \cdot \ln 2} \text{tr} \int_0^1 \mathbf{H}^\dagger \left(\mathbf{I}_{n_r} + \frac{P}{n_t \sigma^2} \mathbf{HH}^\dagger \zeta \right)^{-1} \mathbf{H} d\zeta \qquad (3.40)$$

where tr[] denotes the trace operation.

Using the singular value decomposition, [79], we write $\mathbf{H} = \mathbf{UWV}^\dagger$. Letting $MIN = \min\{n_t, n_r\}$, the matrix $\mathbf{W}$ is an n_t-by-n_r matrix that is all zeroes except that its upper left corner is described as occupied by a MIN by MIN diagonal matrix with the jjth entry w_j. $\mathbf{V}$ and $\mathbf{U}$ are unitary matrices of sizes $n_t \times n_t$ and $n_r \times n_r$, respectively. So, we have

$$\lim_{L \to \infty} C_{SD}^L(\mathbf{H})$$

$$= \frac{P}{\sigma^2 n_t \ln 2} \text{tr} \left[\int_0^1 (\mathbf{VW}^\dagger \mathbf{U}^\dagger)(\mathbf{UU}^\dagger + (P/(\sigma^2 n_t))\mathbf{UWW}^\dagger \mathbf{U}^\dagger \zeta)^{-1} \mathbf{UWV}^\dagger d\zeta \right]$$

$$= \frac{P}{\sigma^2 \cdot n_t \cdot \ln 2} \text{tr} \left\{ \int_0^1 \mathbf{VW}^\dagger \begin{bmatrix} \ddots & & \\ & \left(1 + \frac{P w_j^2 \zeta}{\sigma^2 \cdot n_t}\right)^{-1} & \\ & & \ddots \end{bmatrix} \mathbf{WV}^\dagger d\zeta \right\} \qquad (3.41)$$

where the matrix with inverted diagonal entries depicted above is a diagonal matrix where for $j > MIN$ the term w_j^2 vanishes. Note that the terms w_m^2, $m = 1, 2, \ldots$, and $\min\{n_t, n_r\}$ are the eigenvalues of the matrix $\mathbf{HH}^\dagger$. We denote them as λ_m, $m = 1, 2, \ldots$, and $\min\{n_t, n_r\}$, respectively. Moreover, the trace of a square matrix is unitarily invariant; so, upon carrying out the simple integration, we get

$$\lim_{L \to \infty} C_{SD}^L(\mathbf{H}) = \sum_{m=1}^{MIN} \log_2 \left(1 + \frac{P \lambda_m}{\sigma^2 \cdot n_t} \right) \qquad (3.42)$$

$$= \log_2 \det \left[\mathbf{I}_{n_r} + \frac{P}{\sigma^2 \cdot n_t} \mathbf{HH}^\dagger \right] \qquad (3.43)$$

This is the same as the matrix channel log-det capacity. Thus, we have proved that by knowing the strata rates, log-det capacity is attained in the limit as $L \to \infty$.

3.4.5 Differential Rates of Individual Plies When $L \to \infty$

It is useful to express the accumulation of capacity of SD-BLAST with the progressive peeling of plies. In particular, we want to do this for the asymptote of a large number of plies. We take i to be a fixed fraction, say z of L. Asymptotically, as $L \to \infty$, the differential capacity added by the ith ply approaches the limit

$$\frac{1}{\ln 2} \cdot \sum_{j=1}^{MIN} \frac{P\lambda_j z/(\sigma^2 n_t)}{1 + (P\lambda_j/(\sigma^2 n_t))(1 - z)} dz \tag{3.44}$$

The capacity accumulated through the ith ply is obtained by integrating from 0 to the $Z = i/L$ normalized ply:

$$\sum_{j=1}^{MIN} \log_2 \left[1 + \frac{P\lambda_j Z/(\sigma^2 n_t)}{1 + (P\lambda_j/(\sigma^2 n_t))(1 - Z)} \right] \tag{3.45}$$

As expected, setting $Z = 1$ gives (3.43).

3.4.6 Capacity versus Outage Performance for SD-BLAST and Channel Hardening

In order for SD-BLAST to be optimal, we need to make sure that every ply is coded at the rate seen just below its level of outage mutual information. When the transmitter knows only the statistics of the channel matrix, determining optimal ply rates is an issue of particular interest.

We denote the sum capacity of SD-BLAST using L plies by $C_{SD}^L(\rho, \mathbf{H}) = \sum_{i=1}^{L} C_i^d(\rho, \mathbf{H})$. For comparison purposes, we define the upper bound on the ϵ-capacity of SD-BLAST by

$$\epsilon = \Pr \left[\sum_{i=1}^{n} C_i^d(\rho, \mathbf{H}) \leq C_{SD}^L(\rho, \epsilon) \right] \tag{3.46}$$

where $C_{SD}^L(\rho, \epsilon) = \sum_{i=1}^{L} C_i^d(\rho, \epsilon)$ is the demand on sum capacity at an outage level ϵ and $SNR = \rho$. In SD-BLAST, not only does the condition expressed by (3.46) have to be met, but also each of the plies should meet its corresponding capacity demand $C_i(\rho, \epsilon)$, $\forall i$. The upper bound on the capacity versus outage in (3.46) neglects the possible violations in the ply capacities.

To estimate the actual outage capacity of SD-BLAST, we need to relax (reduce) the demand on the outage capacity. That is, the transmitter needs to make an educated estimate of a lower outage level $(\epsilon - \epsilon')\%$, where $\epsilon > \epsilon' \geq 0$, such that estimates of L ply rates are successful for at least $(100 - \epsilon)\%$ of the channels. We thus define a lower bound on the ϵ-capacity of SD-BLAST as follows:

$$\min_{\epsilon'} \epsilon = \Pr \left[C_i^d(\rho, \mathbf{H}) \leq C_i^d(\rho, \epsilon - \epsilon'), any\ i \right] \tag{3.47}$$

where ϵ' is a penalty imposed on the outage capacity calculation due to the simultaneous optimization of L random capacities. Moreover, a tighter capacity bound can be found by minimizing L slack variables $\epsilon' \geq 0, \forall i$.

Intuitively, we can see that the upper bound on the capacity versus outage performance cannot be achieved by SD-BLAST for a general (n_t, n_r) case because of the randomness of eigenvalues for various realizations of channel matrices. Note, however, that the capacity accumulated through each stratum is related to the channel matrix via its eigenvalues only. Therefore, when there is only one eigenmode, for example $(n_t, 1)$, knowledge of the random capacity C is equivalent to knowledge of the eigenvalue. Thus, the SD-BLAST will approach the capacity versus outage performance upper bound. A similar argument applies for the relatively trivial case $(1, n_r)$. But we are focusing on $n_t > n_r$. Moreover, when $n_t \gg n_r$, the n_t eigenrates become substantially hardened (see [43], [77], and [136]), as they do when both n_t and n_r become large and n_t is a fixed fraction of n_r and we expect a near-optimal performance of capacity versus outage. Assuming that the h_{ij} are i.i.d complex Gaussian $\mathbb{C}\mathcal{N}(0, 1)$:

- For large n_t and fixed n_r ($n_t \gg n_r$ and $n_t \to \infty$), the n_r eigenrates become substantially hardened, and the probability density function, $f(\lambda)$, of eigenvalues of $\mathbf{HH}^\dagger/n_t$ approaches the limiting value $\delta(\lambda - 1)$. Thus, the differential rates accumulated through the $Z = (i/L)$th (normalized) plies in (3.45) become

$$n_r \log_2 \left[1 + \frac{PZ/\sigma^2}{1 + (P/\sigma^2)(1 - Z)} \right] \tag{3.48}$$

  A similar result applies for large n_r and fixed n_t.
- When $n_r = \beta n_t \to \infty$, the eigenvalues λs of $\mathbf{HH}^\dagger/n_t$ tend to become a known continuous density function given by [27]–[31]

$$f(\lambda) = \begin{cases} \frac{1}{\pi}\sqrt{\frac{\beta}{\lambda} - \frac{1}{4}\left(1 + \frac{\beta-1}{\lambda}\right)^2}, & (\sqrt{\beta} - 1)^2 \leq \lambda \leq (\sqrt{\beta} + 1)^2 \\ 0, & \text{otherwise} \end{cases}$$

and the differential rates accumulated through the $Z = (i/L)$ (normalized) plies in (3.45) can be computed in the following closed form:

$$\int_{(\sqrt{\beta}-1)^2}^{(\sqrt{\beta}+1)^2} \log_2 \left[1 + \frac{\rho\lambda Z}{1 + \rho\lambda(1 - Z)} \right] f(\lambda) d\lambda = F(\rho, Z, \beta) \tag{3.49}$$

where $\rho = P/\sigma^2$ and $F(\rho, Z, \beta)$ is independent of eigenvalues. Equation (3.49) shows that the differential rates of plies become independent of eigenvalues as $L \to \infty$.

3.4.7 Capacity versus Outage: The Monte Carlo Method

We will now introduce a general procedure that can be carried out with a Monte Carlo method for computing a lower bound on the capacity versus outage for the stratified space-time architecture. The method is based on the statistical characterization of the channel, assuming that the transmitter knows the channel statistics and nothing more. In the next section, we apply the procedure to an ensemble of random matrix Rayleigh channels and establish its effectiveness in some important cases.

The following is an iterative Monte Carlo method that effectively reduces the penalty ϵ' at a specific outage level $Q\%$-tile. That is, when we consider two different channel populations drawn from the same channel statistical distribution, the outage ply rates estimated must be satisfied for $(100 - Q)\%$ of the channels by proceeding as follows:

1. Estimate the rates (C_{SD}^L and C_i^d, $\forall i$) at a slightly lower outage (relaxed outage), $\bar{q} = (Q - \epsilon_i')\%$, of plies of a randomly chosen fixed channel (h_{ij}) population, Θ_1. The channel population must be large enough to represent the given channel distribution; ϵ' must be maintained as a small but nonnegligible value.

2. Consider next a different channel population, Θ_2, drawn independently from the same channel distribution as in step 1. Calculate the $\overline{Q}\%$ of channels (from the channel population Θ_2) not meeting the weaker demands of ply capacities estimated in step 1, such that we have $\overline{Q} = \Pr[C_i^d(\rho, \mathbf{H}) \leq C_i^d(\rho, \bar{q}), \text{any } i]$.

3. Iterate steps 1 and 2 by perturbing the starting outage Q by ϵ'_{i+1}, where

$$\bar{q} = (Q - \epsilon'_{i+1})\%,$$

until $\overline{Q}\% \simeq Q\%$.

4. (Optional) Steps 1–3 can be repeated by replacing Θ_2 with different channel populations drawn from the same channel distribution to make sure that the desired $Q\%$-tile outage capacity is met more accurately for the estimating procedure.

While the $Q\%$-tile outage capacity is reduced compared to the upper bound in (3.46), in the next section we will see examples where it is fair to say that it can come close.

Remark 4 *As is well known, a matrix channel possesses generalized eigenmodes ([47], [119], [24]). To access these noninterfering modes, the transmitter needs to know the channel matrix. Armed with this knowledge and in accordance with the water-filling procedure of information sending, the transmitter can spatially water pour its available power over the modes to obtain greater capacity than it would if transmitting equal power from each transmit antenna. Each of the summands in (3.43) takes on the role of an eigenrate. We are interested in the eigenrates and the* *rates that the (generalized) eigenmodes can support when transmitting the power $P/(n_t\sigma^2)$ on each transmit antenna. When $n_t > n_r$, the channel has n_r eigenmodes, and the interpretation of (3.43) is that these eigenmodes are being blindly accessed by the transmitter. The wasted power, namely $(n_r - n_t)P/(n_t\sigma^2)$, as well as the inability to water pour, is the price of channel-blind operation.*

3.5 SIMULATIONS ON BLAST FOR THE MATRIX RAYLEIGH CHANNEL

In this section, we present some empirical comparative outage capacity curves of SD-BLAST, V-BLAST, and D-BLAST for a small sampling of (n_t, n_r) systems in an ideal matrix Rayleigh channel. In the examples, the ordered serial interference cancellation decoder used in the V-BLAST system is designed based on the maximum SINR criterion instead of ZF. Moreover, with V-BLAST, using a smaller number of transmitter antennas than n_t can give superior performance than using all available n_t antennas. Therefore, the number of transmitters used in V-BLAST is optimized at 10% outage in all the simulations. The reader is referred to Section 4.4 for a detailed discussion of the capacity calculation of a V-BLAST system.

The average received SNR, ρ', $(\rho' dB = 10 \log_{10} \rho')$ is calibrated as follows. If we make a test measurement that involves transmitting all the available power, P, out of any of the n_t transmit antennas, say the mth, and makes repeated statistically independent channel measurements at any receive antenna, say the jth, then the SNR ρ' is given by

$$\rho' = (P/\sigma^2) \cdot E \; |h_{jm}|^2.$$

In the examples, power P/n_t is transmitted from each of n_t antennas in the transmit array.

3.5.1 Outage Capacity versus SNR at the 10% Outage Level

We now present the outage capacity versus SNR results at the 10% outage level for the (4,1) and (4,2) systems in Figures 3.12 and 3.13, respectively, in the SNR range of $\rho' = -6$ dB to $+24$ dB in 5 dB steps. The SD-BLAST graphs for 1, 2, 4, 8, 16, 32, and 64 strata per layer are shown, along with the outage capacity of the V- and D-BLAST systems. The outage capacity of D-BLAST is the upper bound on the outage capacity performance predicted by theory.

For the (4,1) case, since the transmitter knows the channel statistics, it therefore knows the lower 10%-tile of λ_1. So, in this case, the single eigenrate at the 10% outage level is known to the transmitter. The bit rates per ply are thereby known. Moreover, the estimation procedure was so effective here that the violation counts were negligible for all the n shown. The lower double-dash curve that appears in each figure represents the performance of the upper limit of encoded V-BLAST systems. As expected, V-BLAST exhibits an inferior performance compared to D-BLAST.

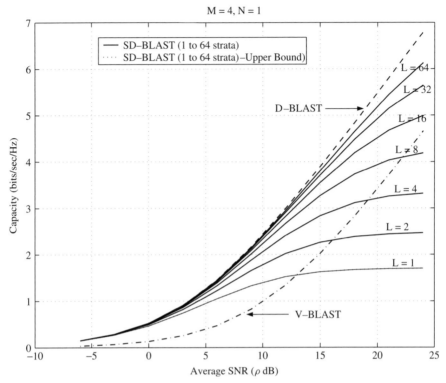

Figure 3.12 (4,1) outage capacity versus average SNR. For (M,1) cases, the estimated 10%-tile rates of SD-BLAST computed using (3.46) and (3.47) effectively coincide.

Figure 3.13 illustrates the (4,2) scenario. With $n_t > n_r > 1$, the problem is now more difficult than the previous one since the transmitter does not know the set of L bit rates per stratum. Estimation of the ply rates is now crucial. We cannot expect to get the same 10%-tiles as we could if the ply rates were known. The aforementioned Monte Carlo estimation technique was used to produce the 10% tile capacities. In Figure 3.13, the degradation of these 10%-tiles from the 10%-tiles upper bound on SD-BLAST is shown, and the degradations are noticeable. The V-BLAST curves exhibit a peak of the deviation. This is because (n_t, n_r) V-BLAST requires that for each SNR reported, the optimum number of transmit antennas less than or equal to n_t was used, and this optimum can change with the SNR value. Such changes in the number of transmitters also occur in the other examples, but in Figure 3.12, there was no noticeable peaking of the slope. Moreover, at a high SNR, the MIMO system is co-antenna interference-limited rather than noise-limited. This is the reason for the saturation of SD-BLAST capacities at a high SNR (see Figures 3.12 and 3.13). The co-antenna interferences are effectively reduced with increasing numbers of strata.

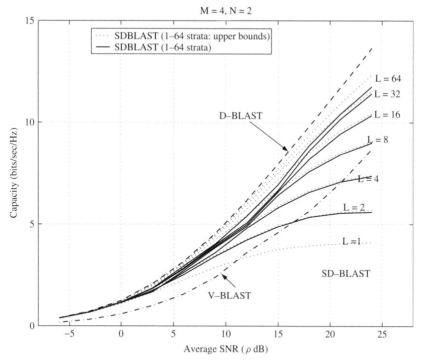

Figure 3.13 (4,2) outage capacity versus average SNR. This case illustrates the degradation of the 10%-tile rates of SD-BLAST compared to the performance upper bound of SD-BLAST.

3.5.2 Capacity Cumulative Density Function Comparison

In this section, we look at the empirical complementary cumulative distribution function (ccdf) of capacity for fixed SNR and various combination of n_t and n_r.

The first three system examples, (4,2), (8,3), and (16, 5), pertain to the important downlink case where the end user has fewer antennas compared to the base(s). Figures 3.14 to 3.16 show the empirical capacity distributions for the first three examples with $n_t > n_r > 1$ at SNR = 18dB. The performance of SD-BLAST is shown for various numbers of strata per layer. The broken thin curves show the capacity distribution of the upper bound on SD-BLAST computed using (3.46). The solid thin curves show the actual capacity distribution of SD-BLAST computed using the aforementioned Monte Carlo method. The figures show the degradation of SD-BLAST from the ccdf curves, which is just about noticeable after $L = 16$. In the (4,2) scenario depicted in Figure 3.14, the excess of n_t over n_r is not as great as in the other two examples. Therefore, as expected, the relative performance is not as good as in the previous two cases. In Figure 3.17, we show the performances of the above three examples at 10 dB SNR. We see again that the greater the excess of n_t over n_r, the better the performance of SD-BLAST.

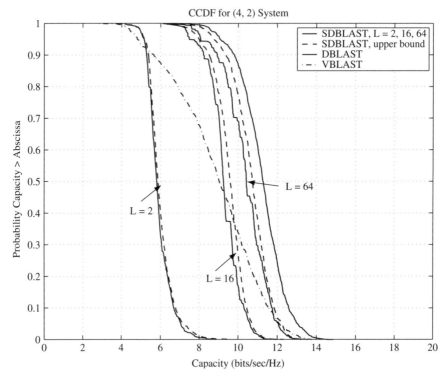

Figure 3.14 Empirical distribution of the ccdf function of the capacity of a (4,2) system. This assumes statistically independent Rayleigh faded paths. Average SNR = 18 dB. This figure illustrates the effectiveness of SD-BLAST compared to V-BLAST for $n_t \gg n_r$.

For all the cases reported here, the performance of D-BLAST (upper bound) and SD-BLAST is significantly worse than that of V-BLAST. It must be stressed that V-BLAST was designed for situations where $n_r > n_t$, so it cannot be expected to exhibit good performance in these situations. Based on the results reported here, we may conclude that SD-BLAST is a good choice for $n_t > n_r$.

In Figure 3.18, we show the performance of SD-BLAST in a (2,4) system for 16 strata along with that of V-BLAST and D-BLAST for SNR = 10 dB and 18 dB. Although the performance of D-BLAST and SD-BLAST is worse than that of V-BLAST for the (2,4) system (but not significant as in the $n_t > n_r$ cases), V-BLAST may be considered a better choice for $n_t < n_r$ than SD-BLAST due to practical design considerations.

3.6 MULTIRATE LAYERED SPACE-TIME ARCHITECTURE

In this section, we introduce another BLAST architecture called multirate layered space-time (MLST) architecture. The proposed framework can be viewed as a class

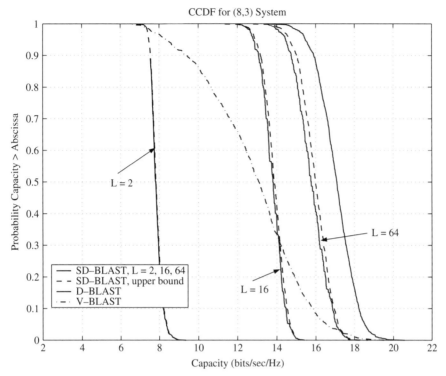

Figure 3.15 Empirical distribution of the ccdf function of the capacity of an (8,3) system. This assumes statistically independent Rayleigh faded paths. Average SNR = 18 dB. This figure illustrates the effectiveness of SD-BLAST compared to V-BLAST for $n_t \gg n_r$.

of diagonal layered space-time coded systems with each of the layers encoded independently, with different rates subject to equal per-layer outage probabilities. The key challenge involved in this architecture is estimating the per-layer rates. Note that in a horizontally encoded V-BLAST system, equal rate codes are used for each layer.

The idea of successive decoding, which involves decoding users sequentially in a given order, was initially proposed in an information-theoretic study of scalar output Gaussian multiple access channels ([28], [153] and [165]). In this scheme, the first user is decoded by treating the interference from other users as noise. The decoded transmitted codeword of the first user is then subtracted from the received data, the second user is decoded by treating the interference from the remaining users as noise, and so on. It is well known that, in theory, minimum mean-square error (MMSE)–based successive decoding and interference cancellation type of receivers of multiple-access (multiuser communications) and MIMO channels achieve the total Shannon capacity, provided that, for each layer (user), optimal coding rates are known ([6], [44]). Thus, the channel capacity can be achieved by independent encoding and successive decoding procedures using 1-D

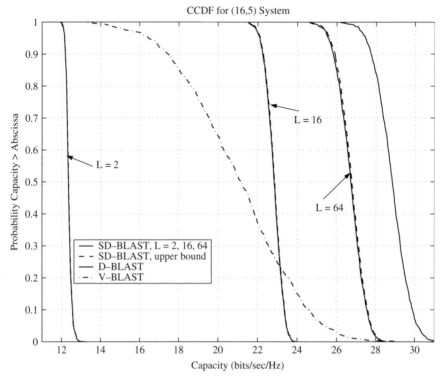

Figure 3.16 Empirical distribution of the ccdf function of the capacity of a (16,5) system. This assumes statistically independent Rayleigh faded paths. Average SNR = 18 dB. This figure illustrates the effectiveness of SD-BLAST compared to V-BLAST for $n_t \gg n_r$.

channel codes. However, for this scheme to perform optimally, the optimal rate of each layer must be known at the transmitter. In this section, we discuss such an architecture, the MLST architecture and the important issue of identifying the optimal rates for the layers to achieve capacity. The term layer comes from the BLAST architectures.

Consider then an i.i.d. Rayleigh fading MIMO system that has n_t transmit and n_r receive antennas with additive Gaussian noise at each of the receive antennas. This assumption is reasonable for indoor and urban areas with well-spaced multiple antenna elements. In complex baseband representation, the system can be written, considering T symbol periods, as follows:

$$\mathbf{Y} = \mathbf{HX} + \mathbf{V} \tag{3.50}$$

where $\mathbf{X} \in \mathbb{C}^{n_t \times T}$ and its ith row corresponds to the transmitted signal at the ith antenna, $i = 1, \ldots, n_t$. Similarly, $\mathbf{Y} \in \mathbb{C}^{n_r \times T}$ and its ith row corresponds to the received signal at the ith receive antenna, $i = 1, \ldots, n_r$, and $\mathbf{H} = [\mathbf{h}_1, \mathbf{h}_2, \ldots, \mathbf{h}_{n_t}]$ is the $n_r \times n_t$ channel matrix with $\mathbf{h}_j = [h_{1j}, \ldots, h_{Nj}]^T$. The propagation gains

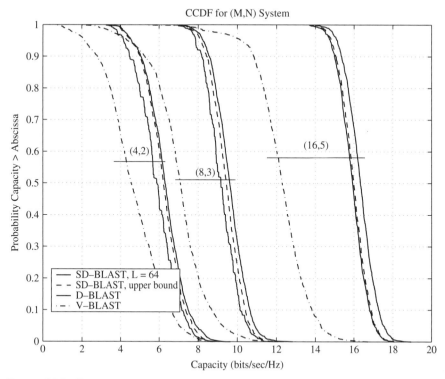

Figure 3.17 Empirical distribution of the ccdf function of the capacity of (4,2), (8,3), and (16,5) systems. This assumes statistically independent Rayleigh faded paths. Average SNR = 10 dB. This figure illustrates the improved performance of SD-BLAST with excess transmit antennas for $n_t \gg n_r$.

from the jth transmit antenna to the ith receive antenna h_{ij} are treated as i.i.d. complex Gaussian variables with zero mean and unit variance, i.e., $\mathbf{H} \sim \mathcal{CN}(0, I_{n_r} \otimes I_{n_t})$. Here we read the symbol "$\sim$" as "is distributed as," $\mathcal{CN}$ denotes the complex normal distribution, and $\otimes$ denotes the Kronecker product. The additive white Gaussian noise $\mathbf{V} \in \mathbb{C}^{n_r \times T}$ has i.i.d. entries, i.e., $v_{ij} \sim \mathcal{CN}(0, \sigma^2)$, where the noise component has zero mean and variance σ^2. We normalize the power to satisfy the condition $\mathcal{E}\left\{ \sum_{i=1}^{n_t} \sum_{j=1}^{T} |x_{ij}|^2 \right\} = PT$, where P is the total power transmitted per symbol interval. The matrix channel $\mathbf{H}$ is quasi-static and remains constant over the T symbol periods or the duration of a packet.

3.6.1 Encoder-Decoder Structure

The proposed transmitter is illustrated in Figure 3.19. In this structure, a primitive data stream is demultiplexed into n_t substreams, which are encoded independently using 1-D channel codes with varying code rates $R_i, i = 1, 2, \ldots, n_t$ to form coded

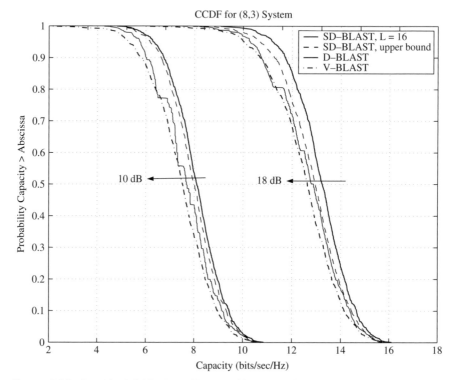

Figure 3.18 Empirical distribution of the ccdf function of the capacity of a (2,4) system. This assumes statistically independent Rayleigh faded paths. Average SNR = 10 dB and 18 dB. This figure shows that simple V-BLAST is competitive with SD-BLAST for $n_r \gg n_t$.

layers $A_i, i = 1, 2, \ldots, n_t$. In the multirate encoding and successive decoding structure, for quasi-static or slow-fading channel environments, a diagonal space-time interleaver can be introduced that cycles each of the independently encoded layers over the transmit antennas before transmitting, as shown in Figure 3.19(a). Figure 3.19(b) shows the diagonal space-time interleaver (DSTI), where each independent layer is cycled through all the transmit antennas. The diagonal interleaving allows full access to the channel spatial diversity in the absence of the other layers. The transmitter structure in Figure 3.19(a) without the DSTI can be regarded as the horizontal layered space-time (HLST) coding described in Section 4.4.3, where we have $A_i = X_i, \forall i$.

At the receiving end, the decoupling process for each of the n_t layers involves a combination of nulling (minimizing) the interference from yet undetected signals and canceling out the interference from already detected signals. Thus, the SDIC receiver involves a two-step processing procedure: (1) feedforward (FF) and (2) feedback (FB) processing, which are parameterized by a corresponding set of FF and FB filters, respectively. The FF filters estimate the substreams based on the modified received signal, whereas the FB filters cancel the interference from

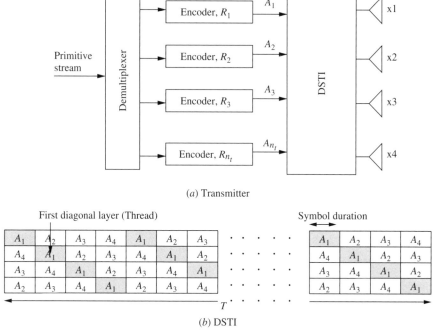

(a) Transmitter

(b) DSTI

Figure 3.19 (a) Transmitter for quasi-static fading channels for a four-transmit-antenna system; (b) DSTI.

the past estimated signals, as shown in Figure 3.20. Without loss of generality, we assume that the layers are decoded in an increasing order of indices.

Note that due to DSTI, each layer is sent simultaneously by all transmit antennas. Therefore, we need to define a set of n_t FF and FB filters for each layer. Let the FF filters be described by the set of n_r-dimensional vectors $\mathbf{f}_k^i$, $i, k = 1, 2, \ldots, n_t$, where i and k are the antenna and layer indices, respectively. Similarly, the FB filters are defined by the set of n_r-dimensional vectors $\mathbf{b}_k^i$, $k = 1, 2, \ldots, n_t - 1$, $i = 1, 2, \ldots, n_t$. When decoding the kth layer, the interference from the already decoded layers is removed by subtracting a linear combination of the decoded[3] stream, $\hat{A}_{k-1}$, and the kth layer FB filters from $\mathbf{f}_k^i \mathbf{Y}_{k-1}$. In other words, we have

$$\mathbf{Y}_1 = \mathbf{f}_1^i \mathbf{Y}$$

$$\mathbf{Y}_2 = \mathbf{f}_2^i \mathbf{Y}_1 - \mathbf{b}_1^i \hat{A}_1$$

$$\vdots = \vdots$$

$$\mathbf{Y}_k = \mathbf{f}_k^i \mathbf{Y}_{k-1} - \mathbf{b}_{k-1}^i \hat{A}_{k-1}$$

---

[3]Interference nulling and cancellation processes are followed by channel decoding, reencoding, and remodulation.

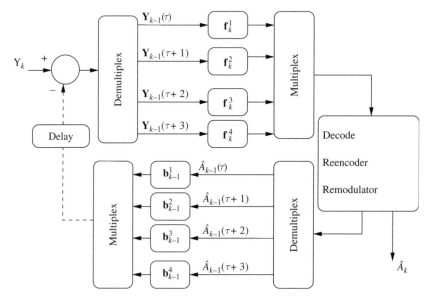

Figure 3.20 Successive decoding and interference cancellation receiver showing the kth decoding layer. Here $(\tau \bmod n_t) = 0$ with $n_t = 4$.

The inner product of the resulting modified received vector $(\mathbf{Y}_k)$ and the $(k+1)$th layer FF filters gives an estimate of the kth decoded signal stream, as shown in Figure 3.20. We assume that all the previous layers are decoded error-free so that error propagation does not occur.

Each layer is cycled through all the transmit antennas according to the DSTI, as illustrated in Figure 3.19(b). Therefore, the effective channel gain and the interferences corresponding to different antennas of a layer are time-varying, which will effectively increase the per-layer rates for larger numbers of transmit antennas. Thus, we can denote the effective channel matrix of antenna i as a column-wise shifted version of the channel matrix $\mathbf{H}$ given by

$$\mathbf{C}^i = [\mathbf{h}_i, \mathbf{h}_{i-1}, \ldots, \mathbf{h}_2, \mathbf{h}_1, \mathbf{h}_{n_t}, \mathbf{h}_{n_t-1}, \ldots, \mathbf{h}_{i+1}] := [\mathbf{c}_1^i, \ldots, \mathbf{c}_{n_t}^i] \forall i$$

The effective channel matrices are used to find the corresponding FF and FB filters.

3.6.2 Optimal Filters — with DSTI

Here we consider the transceiver structure with DSTI, as shown in Figure 3.19. The optimal FF and FB filters that maximize the MI of ith transmit antenna of layer k can be found by maximizing the channel capacity [153]:

$$\mathcal{I}_k^i = \max_{\mathbf{f}_k^i, \mathbf{b}_k^i} \log(1 + \theta \Upsilon_k^i) \tag{3.51}$$

where $\theta = \rho/n_t$, $\rho = P/\sigma^2$ is the average SNR per receive antenna and $\theta \Upsilon_k^i$ is the signal-to-interference ratio (SIR). To find the optimum filters, we can alternatively maximize the SIR and obtain the FB and FF filters as

$$\mathbf{b}_k^i = \mathbf{c}_k^i, \quad k = 1, \ldots, n_t - 1$$

$$\mathbf{f}_k^i = \alpha \left[\mathbf{C}_k^i (\mathbf{C}_k^i)^\dagger + \frac{1}{\theta} \mathbf{I} \right]^{-1} \mathbf{c}_k^i, \quad k = 1, \ldots, n_t \qquad (3.52)$$

where $\mathbf{C}_k^i = [\mathbf{c}_{k+1}^i, \mathbf{c}_{k+2}^i, \ldots, \mathbf{c}_{n_t}^i]$ is an $n_r \times (n_t - k)$ matrix constructed from the channel vectors of the $(n_t - k)$ interfering signals at the kth layer and $\alpha = 1 + 1/\theta$. The maximum SIR of the kth layer signal is given by $\theta \Upsilon_k^i$, where

$$\Upsilon_k^i = (\mathbf{c}_k^i)^\dagger \left[\mathbf{I} + \theta \mathbf{C}_k^i (\mathbf{C}_k^i)^\dagger \right]^{-1} \mathbf{c}_k^i \qquad (3.53)$$

3.6.3 Optimal Filters — without DSTI

When the transmitter structure does not use the DSTI (i.e, in Figure 3.19), we omit the DSTI component; that is, layer i is transmitted by the ith antenna only). The optimal FF and FB filters of the multirate scheme are given in [128] by

$$\mathbf{b}_k = \mathbf{h}_k, \quad k = 1, \ldots, n_t - 1$$

$$\mathbf{f}_k = \alpha \left[\mathbf{H}_k \mathbf{H}_k^\dagger + \frac{1}{\theta} \mathbf{I} \right]^{-1} \mathbf{h}_k, \quad k = 1, \ldots, n_t \qquad (3.54)$$

where $\mathbf{H}_k = [\mathbf{h}_{k+1}, \mathbf{h}_{k+2}, \ldots, \mathbf{h}_{n_t}]$ is an $n_r \times (n_t - k)$ matrix constructed from channel vectors of the $(n_t - k)$ interfering signals at the kth layer; see [153] for details. In this case, the SIR of the kth antenna (or layer) signal is given by $\theta \Upsilon_k$, where

$$\Upsilon_k = \mathbf{h}_k^\dagger [\mathbf{I} + \theta \mathbf{H}_k \mathbf{H}_k^\dagger]^{-1} \mathbf{h}_k \qquad (3.55)$$

The solution (3.54) can also be obtained by using the MMSE FF filters. Moreover, using Cholesky decomposition [57], all of these MMSE FF filters can be written as

$$[\mathbf{f}_1, \ldots, \mathbf{f}_{n_t}] := \mathbf{F}_{\text{mmse}} = (\mathbf{L}_{\text{mmse}}^\dagger)^{-1} \mathbf{H}^\dagger \quad \text{with} \quad \mathbf{H}^\dagger \mathbf{H} + \mathbf{I} = \mathbf{L}_{\text{mmse}}^\dagger \mathbf{D}_{\text{mmse}} \mathbf{L}_{\text{mmse}} \qquad (3.56)$$

Here $\mathbf{L}_{\text{mmse}}$ is a lower triangular matrix with unity diagonal elements and $\mathbf{D}_{\text{mmse}}$ is a diagonal matrix with elements d_i^{mmse}, $i = 1, \ldots, n_t$; see [61]. The SIR of the kth antenna (or layer) signal is now given by $\theta \Upsilon_k$, where $\Upsilon_k = d_k^{\text{mmse}} - 1$. If we use the ZF or nulling FF filters, the nulling FF filters can also be written in matrix form, similar to solution (3.56), as

$$\mathbf{F}_\nu = (\mathbf{L}_\nu^\dagger)^{-1} \mathbf{H}^\dagger \quad \text{with} \quad \mathbf{H}^\dagger \mathbf{H} = \mathbf{L}_\nu^\dagger \mathbf{D}_\nu \mathbf{L}_\nu \qquad (3.57)$$

Here the SIR of the kth antenna signal is given by $\theta \Upsilon_k$, where $\Upsilon_k = d_k^v$ is a chi-square random variable with $2(n_r - n_t + k)$ degrees of freedom and $d_1^v, \ldots, d_{n_t}^v$ are independent. This follows from the fact that an $n_r \times n_t$ complex Gaussian random channel matrix $\mathbf{H}$ is distributed as $\mathbf{H} \sim \mathcal{C}N(0, I_{n_r} \otimes I_{n_t})$ with mean $\mathcal{E}\{\mathbf{H}\} = 0$ and covariance $\text{cov}\{\mathbf{H}\} = I_{n_r} \otimes I_{n_t}$. Then the matrix $\mathbf{W} = \mathbf{H}^\dagger \mathbf{H}$ is called a complex central Wishart matrix [2] and its distribution is denoted by $\mathcal{C}W_{n_t}(n_r, I_{n_t})$. Let $\mathbf{W} = \mathbf{H}^\dagger \mathbf{H} = \mathbf{L}_v^\dagger \mathbf{D}_v \mathbf{L}_v$. Then the matrix elements l_{ij}^v, $1 \le j < i \le n_t$ and $d_1^v, \ldots, d_{n_t}^v$ are independently distributed with $l_{ij}^v \sim \mathcal{C}N(0,1)$, $1 \le j < i \le n_t$ and d_k^v is a chi-square random variable with $2(n_r - n_t + k)$ degrees of freedom [58].

If the transmitter structure uses the DSTI, then we need to use the column-wise shifted version of the channel matrix $\mathbf{H}$ (i.e., $\mathbf{C}^i \, \forall i$) to obtain formulas similar to (3.56) and (3.57) for the FF filter vectors.

3.7 OUTAGE CAPACITY

In a quasi-static channel setting, the matrix channel $\mathbf{H}$ is chosen randomly according to some probability distribution and kept approximately constant for the duration of transmission. When, however, the channel is used from time to time, its transfer characteristic is constant for the duration of its communication burst, yet the channel can change significantly from one use to the next. In this case, we may speak of a trade-off between outage probability and outage capacity. Specifically, we will control this outage probability to a prescribed tolerable value ϵ. The total rate achieved by the system at the outage probability ϵ is denoted by R_ϵ.

The outage capacity is defined in the limit of large block sizes, i.e., as $T \to \infty$. Nevertheless, the outage capacity may remain effective in predicting the capacity behavior of moderately coded systems. The capacity a specific channel can support is given by the following expression [149]:

$$\mathcal{I} = \log \det (\mathbf{I} + \theta \mathbf{H}^\dagger \mathbf{H}) \tag{3.58}$$

When the statistics of $\mathbf{H}$ are known at the transmitter, the capacity versus outage performance is defined by the ϵ-capacity and the outage probability ϵ is defined in probabilistic terms as

$$\epsilon(\mathcal{I}) = 1 - \Pr[\mathcal{I} > R_\epsilon] \tag{3.59}$$

Here, at a certain SNR, once a rate of transmission is chosen, there will be a nonnegligible probability ϵ that the value of the chosen transmission rate R_ϵ exceeds the actual MI, in which case communications fails, i.e. an outage occurs. However, a properly chosen channel code is capable of supporting all but a small fraction of channel realizations. The unique inverse function of $\epsilon(\mathcal{I})$ is called the outage MI (or outage capacity) and is denoted by $\text{out}(\epsilon_{\text{out}})$ when $\epsilon_{\text{out}} = \epsilon(\mathcal{I}_{\text{out}})$. In other words, we define $\text{out}(\epsilon_{\text{out}})$ such that

$$\mathcal{I}_{\text{out}} = \text{out}(\epsilon_{\text{out}}) \quad \text{where} \quad \epsilon_{\text{out}} = \epsilon(\mathcal{I}_{\text{out}})$$

Here $\mathcal{I}_{\text{out}} = \text{out}(\epsilon_{\text{out}})$ means that, in any instantiation of **H** from the ensemble, we obtain an MI $\mathcal{I}$ less than $\mathcal{I}_{\text{out}}$ with probability ϵ_{out} (outage probability).

3.7.1 Per-Layer Rates — without DSTI

First, we consider the transmitter structure in Figure 3.19(a) without DSTI. The instantaneous per-layer rate is defined as

$$\mathcal{I}_k = \log(1 + \theta \Upsilon_k), \quad k = 1, 2, \ldots, n_t \tag{3.60}$$

where $\theta \Upsilon_k$ is the SIR. Since the outage probability $\epsilon_k(\mathcal{I}_k)$ is strictly increasing (or monotonically increasing) function in the per-layer rate $\mathcal{I}_k$, the maximum sum rate $\sum_{k=1}^{n_t} R_{k,\epsilon}$ is achieved at equal per-layer outage probabilities, that is,

$$\epsilon_1(R_{1,\epsilon}) = \epsilon_2(R_{2,\epsilon}) = \cdots = \epsilon_{n_t}(R_{n_t,\epsilon}) = \epsilon$$

In other words, this solution will ensure a maximum sum rate subject to an upper bound on the per-layer outage probability ϵ (see [167]), that is,

$$\max \left\{ \sum_{k=1}^{n_t} \mathcal{I}_k : \quad \epsilon_k(\mathcal{I}_k) \leq \epsilon \, \forall \, k \right\} \tag{3.61}$$

Since $\epsilon_k(\mathcal{I}_k)$ in (3.61) is continuous and increasing (or strictly increasing) in per-layer rate $\mathcal{I}_k$, (3.61) may be replaced by

$$\max \left\{ \sum_{k=1}^{n_t} \mathcal{I}_k : \quad \epsilon_k(\mathcal{I}_k) = \epsilon \, \forall \, k \right\}$$

Therefore, the maximum sum rate $\sum_{k=1}^{n_t} R_{k,\epsilon}$ is achieved at $\epsilon_1(R_{1,\epsilon}) = \epsilon_2(R_{2,\epsilon}) = \cdots = \epsilon_{n_t}(R_{n_t,\epsilon}) = \epsilon$.

Moreover, let $R_{k,\epsilon}$ denote the maximum rate of the kth decoded layer; for this particular channel we can say that the channel is in outage if there exists at least one k for which $\mathcal{I}_k \leq R_{k,\epsilon}$. The event $\mathcal{I}_k > R_{k,\epsilon}$ implies that the effective channel seen by the kth layer, for a given choice of channel and power distribution, can support the rate $R_{k,\epsilon}$. For arbitrarily large (random) block codes, using the Shannon coding theorem, there exist good codes that can achieve arbitrarily low error probabilities. Thus, we can reencode each layer of symbols to be error-free and assume perfect cancellation of detected error-free signals (i.e., no error propagation). Then the outage probability may be defined as

$$\epsilon = 1 - \Pr(\mathcal{I}_1 > R_{1,\epsilon}, \ldots, \mathcal{I}_{n_t} > R_{n_t,\epsilon}) \tag{3.62}$$

where the per-layer outage probability ϵ is the same for all layers. If per-layer rates $\mathcal{I}_1, \ldots, \mathcal{I}_{n_t}$ are independent, then we have

$$\epsilon = 1 - \prod_{k=1}^{n_t} \Pr(\mathcal{I}_k > R_{k,\epsilon}) = 1 - (1 - \epsilon)^{n_t} \tag{3.63}$$

Otherwise, using Bonferroni inequality (or Boole's inequality [76]), we obtain

$$\epsilon \le n_t - \sum_{k=1}^{n_t} \Pr(\mathcal{I}_k > R_{k,\epsilon}) = n_t \epsilon \tag{3.64}$$

The maximum sum rate achieved by this system at probability ϵ is $\sum_{k=1}^{n_t} R_{k,\epsilon}$.

The Rayleigh fading channel is assumed to be i.i.d. (i.e., $\mathbf{H} \sim \mathcal{CN}(0, I_{n_r} \otimes I_{n_t})$). Therefore, when we use the ZF FF filter, the SIRs $\theta \Upsilon_1, \ldots, \theta \Upsilon_{n_t}$ are independent ([61] and [167]) and the per-layer rates $\mathcal{I}_1, \ldots, \mathcal{I}_{n_t}$ are likewise independent. At a high SNR, the per-layer rates and outage probability obtained for the ZF receiver can be readily extended for the MMSE receiver. However, in general, it is difficult to analyze the statistical properties of the per-layer rates when we use an MMSE FF filter.

Next, we consider the ZF or nulling FF filter and evaluate the outage probability (3.62). The following proposition is required in what follows.

Proposition 1 *Let z be distributed as a chi-square random variable with $2n$ degrees of freedom, and let the probability density function be denoted by $f(z)$, i.e., $z \sim \mathcal{X}_{2n}^2$. Then the probability density function of $\mathcal{I} = \log(1 + \theta z)$ is given by*

$$q(\mathcal{I}) = \frac{e^{\mathcal{I}}}{\theta} f\left(\frac{e^{\mathcal{I}} - 1}{\theta}\right) \tag{3.65}$$

and the ccdf is given by

$$\Pr(\mathcal{I} > R) = \frac{\Gamma(n, (e^R - 1)/\theta)}{\Gamma(n)} = e^{-(e^R - 1)/\theta} \sum_{i=0}^{n-1} \frac{(e^R - 1)^i}{\theta^i i!} \tag{3.66}$$

where the upper incomplete gamma function is given by $\Gamma(n, a) = \int_a^\infty t^{n-1} e^{-t}\, dt$, the complete gamma function is given by $\Gamma(n) = \Gamma(n, 0)$, and $\theta = \rho/n_t$.

Proof. The density of the MI can be written as

$$q(\mathcal{I}) = \int \delta(\mathcal{I} - \log(1 + \theta z)) f(z)\, dz$$

$$= \frac{e^{\mathcal{I}}}{\theta} \int \delta\left(z - \frac{e^{\mathcal{I}} - 1}{\theta}\right) f(z)\, dz$$

$$= \frac{e^{\mathcal{I}}}{\theta} f\left(\frac{e^{\mathcal{I}} - 1}{\theta}\right) \tag{3.67}$$

where $\delta(\cdot)$ is a delta function. The ccdf is given by

$$\Pr(\mathcal{I} > R) = \int_R^\infty g(\mathcal{I}) \, d\mathcal{I}$$

$$= \frac{1}{\Gamma(n)} \int_R^\infty \frac{e^{\mathcal{I}}}{\theta} \left(\frac{e^{\mathcal{I}} - 1}{\theta} \right)^{n-1} e^{-(e^{\mathcal{I}}-1)/\theta} d\mathcal{I}$$

$$= \frac{1}{\Gamma(n)} \int_{(e^R-1)/\theta}^\infty y^{n-1} e^{-y} \, dy$$

$$= \frac{\Gamma(n, (e^R - 1)/\theta)}{\Gamma(n)} \tag{3.68}$$

The result follows. $\qquad\qquad\qquad\qquad\qquad\qquad\qquad\qquad\qquad\square$

The ccdf of the MI given in (3.68) is similar to the chi-square ccdf. Therefore, we can easily understand the statistical behavior of the random MI $\mathcal{I}_k$. From (3.62), if we use the ZF FF filter and $\mathcal{I}_1, \ldots, \mathcal{I}_{n_t}$ are independent, then the outage probability is given by

$$\epsilon_v = 1 - \prod_{k=1}^{n_t} \left[e^{-(e^{R_{k,\epsilon}}-1)/\theta} \sum_{i=0}^{n_r-n_t+k-1} \frac{(e^{R_{k,\epsilon}} - 1)^i}{\theta^i i!} \right] \tag{3.69}$$

3.7.2 Per-Layer Rates — with DSTI

When we use the transmitter structure in Figure 3.19(a) with a DSTI and an MMSE FF filter, the instantaneous per-layer rate can be defined as

$$\mathcal{I}_k^d = \frac{1}{n_t} \sum_{i=1}^{n_t} \log(1 + \theta \Upsilon_k^i), \quad k = 1, 2, \ldots, n_t \tag{3.70}$$

where $\theta \Upsilon_k^i$ is the SIR of the kth layer signal sent by antenna i and Υ_k^i is given in (6.30). The per-layer rate is obtained by averaging over all possible channels. Therefore, the density of the per-layer MI is concentrated around its mean[4] (variance is reduced by $1/n_t$). This implies that as n_t increases, in an asymptotic sense the outage and ergodic capacities approach the same value. The outage probability for this case is given by

$$\epsilon_{\text{mmse}} = 1 - \Pr(\mathcal{I}_1^d > R_{1,\epsilon}^d, \ldots, \mathcal{I}_{n_t}^d > R_{n_t,\epsilon}^d) \tag{3.71}$$

If $\mathcal{I}_1^d, \ldots, \mathcal{I}_{n_t}^d$ are dependent, then using Bonferroni inequality, we have

---

[4]If Gaussian random variables $X_1, \ldots, X_n$ are distributed as $X_i \sim \backslash\nabla(\mu, \sigma^2) \forall i$, then the mean $\overline{X}$ is distributed as $\overline{X} = \frac{1}{n} \sum_{i=1}^n X_i \sim \backslash\nabla(\mu, \sigma^2/n)$.

$$\epsilon_{\text{mmse}} \leq n_t - \sum_{k=1}^{n_t} \Pr(\mathcal{I}_k^d > R_{k,\epsilon}^d) = n_t \epsilon.$$

The maximum sum rate achieved by this system at probability ϵ_{mmse} is $\sum_{k=1}^{n_t} R_{k,\epsilon}^d$. These maximum per-layer rates $R_{1,\epsilon}^d, \ldots, R_{n_t,\epsilon}^d$ are obtained at equal per-layer outage probability ϵ.

If we use the ZF FF filter, then Υ_k^i is a chi-square random variable with $2(n_r - n_t + k)$ degrees of freedom and $\mathcal{I}_1^d, \ldots, \mathcal{I}_{n_t}^d$ are independent. Hence, we have $\epsilon_v = 1 - \prod_{k=1}^{n_t} \Pr(\mathcal{I}_k^d > R_{k,\epsilon}^d) = 1 - (1 - \epsilon)^{n_t}$ and we can easily compute this outage probability. The following theorem gives the probability density function of $\mathcal{I}_k^d$ when we use ZF.

Theorem 1 *Let $z_1, \ldots, z_{n_t}$ be distributed as chi-square random variables with $2n$ degrees of freedom, and let the joint probability density function of z_i's be denoted by $g(z_1, \ldots, z_{n_t})$. Then the probability density function of $\mathcal{I} = 1/n_t \sum_{i=1}^{n_t} \log(1 + \theta z_i)$ is given by*

$$q(\mathcal{I}) = \frac{n_t e^{n_t \mathcal{I}}}{\theta^{n_t}} \int_1^{e^{n_t \mathcal{I}}} \int_1^{x_1} \cdots \int_1^{x_{n_t-2}}$$

$$\times g\left(\frac{e^{n_t \mathcal{I}} - x_1}{\theta x_1}, \frac{x_1 - x_2}{\theta x_2}, \ldots, \frac{x_{n_t-2} - x_{n_t-1}}{\theta x_{n_t-1}}, \frac{x_{n_t-1} - 1}{\theta}\right) \prod_{i=1}^{n_t-1} \frac{1}{x_i} dx_i \quad (3.72)$$

where $\theta = \rho/n_t$.

Proof. The density of the MI can be written as

$$q(\mathcal{I}) = \int \cdots \int \delta\left(\mathcal{I} - \frac{1}{n_t} \log\left[\prod_{i=1}^{n_t}(1 + \theta z_i)\right]\right) g(z_1, \ldots, z_{n_t}) dz_1 \cdots dz_{n_t}$$

$$= n_t e^{n_t \mathcal{I}} \int \cdots \int \delta\left(e^{n_t \mathcal{I}} - \prod_{i=1}^{n_t}(1 + \theta z_i)\right) g(z_1, \ldots, z_{n_t}) dz_1 \cdots dz_{n_t}$$

$$= n_t e^{n_t \mathcal{I}} \int \cdots \int \delta\left(e^{n_t \mathcal{I}} - (1 + \theta z_1)x_1\right) \delta(x_1 - (1 + \theta z_2)x_2) \cdots \delta$$

$$(x_{n_t-1} - (1 + \theta z_{n_t}))g(z_1, \ldots, z_{n_t}) dz_1 \cdots dz_{n_t} dx_1 \cdots dx_{n_t-1}$$

$$= M e^{n_t \mathcal{I}} \int \cdots \int \frac{1}{\theta x_1} \delta\left(z_1 - \frac{e^{n_t \mathcal{I}} - x_1}{\theta x_1}\right) \frac{1}{\theta x_2} \delta\left(z_2 - \frac{x_1 - x_2}{\theta x_2}\right) \cdots \frac{1}{\theta} \delta$$

$$\left(z_{n_t} - \frac{x_{n_t-1} - 1}{\theta}\right) g(z_1, \ldots, z_{n_t}) dz_1 \cdots dz_{n_t} dx_1 \cdots dx_{n_t-1}$$

The result follows. $\qquad\square$

We can compute the outage probability using Theorem 1. For example, if $n_t = n_r = 2$, then the probability density function of $\mathcal{I}_k^d$ is given by

$$q(\mathcal{I}) = \frac{2e^{2\mathcal{I}}}{\theta^2} \int_1^{e^{2\mathcal{I}}} \frac{1}{x} g\left(\frac{e^{2\mathcal{I}} - x}{\theta x}, \frac{x - 1}{\theta}\right) dx \qquad (3.73)$$

where $g(z_1, z_2)$ is the joint density of $z_1, z_2 \sim \mathcal{X}_{2n}^2$ with $n = n_r - n_t + k$. Let the density of z_i be denoted by $f(z_i)$. If z_1 and z_2 are independent, then $g(z_1, z_2) = \prod_{i=1}^{2} f(z_i)$. Figure 3.21 shows the probability density functions of the MI and the outage capacities. When we use DSTI, the density of the per-layer MI tends to concentrate around its mean (i.e., variance is reduced by $1/n_t$). Hence, it achieves more outage capacity at low values of outage probability, which is the case of interest in practical schemes. If we increase n_t, then asymptotically the outage and ergodic capacities coincide [85]. For example, the 10% outage capacity of the first layer of the multirate coded system with DSTI is 1.19 bits, whereas for the system without DSTI it is 0.88 bits.

From the per-layer MI densities (3.65) and (3.72), we can estimate the optimum per-layer rates $R_{k,\epsilon}$ at equal outage probability $\epsilon_k(R_{k,\epsilon})$, i.e.,

$$\epsilon_k(R_{k,\epsilon}) = \int_0^{R_{k,\epsilon}} q(\mathcal{I})d\mathcal{I} := \epsilon \qquad (3.74)$$

The per-layer MI densities are difficult to obtain analytically if we use MMSE FF filter. In this case, we obtain the per-layer rates using the simulation procedure (or Monte Carlo method). First, we generate sufficiently large numbers of i.i.d. random $n_r \times n_t$ channel matrices at different SNRs. Then we estimate the n_t per-layer rates (i.e., $\mathcal{I}_k^d$, $\forall k$) according to (3.70) and the total outage channel capacity (3.58). Finally, we find the optimum per-layer rates at equal per-layer outage probability ϵ and the corresponding sum rate at an outage level (say, ϵ_{mmse}) satisfying the conditions given in (3.71). The per-layer rates for multirates without DSTI are calculated using the same procedure.

3.8 SIMULATION RESULTS

As mentioned previously, we empirically estimate the optimum per-layer rates of different schemes using randomly generated $n_r \times n_t$ channel matrices. Then we compare the maximum achievable information rate versus SNR for various antenna configurations for the following: (1) ergodic capacity, (2) outage capacity $R_{1\%}$, (3) sum capacity $\sum_{k=1}^{n_t} R_{k,1/n_t\%}^d$ with DSTI, (4) sum-capacity $\sum_{k=1}^{n_t} R_{k,1/n_t\%}$ without DSTI, and (5) 1% outage capacity of the V-BLAST; see [44]. We compute the $\frac{1}{n_t}\%$ per-layer outage capacity, which will give at most 1% sum-rate, i.e., $\epsilon_{mmse} \leq 0.01$.

Figures 3.22(a)–(D) illustrate capacity versus SNR for $n_t = n_r = 2, 4, 6$, and 16. From the figures, we observe the following:

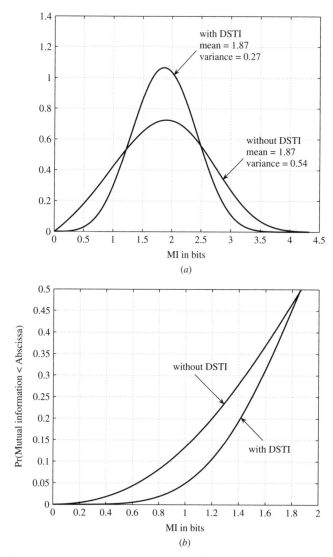

Figure 3.21 (a) Probability density functions of the per-layer MI for $M = N = 2$, $k = 2$, and SNR = 5 dB; (b) per-layer outage capacities for $M = N = 2$, $k = 2$, and SNR = 5 dB.

- When we use DSTI at the transmitter, the 1% outage capacity is higher than that of the scheme without DSTI.
- For large numbers of transmit antennas, the scheme with DSTI approaches the capacity.
- The performance of a V-BLAST is better than the scheme without DSTI; see [167] for similar results.

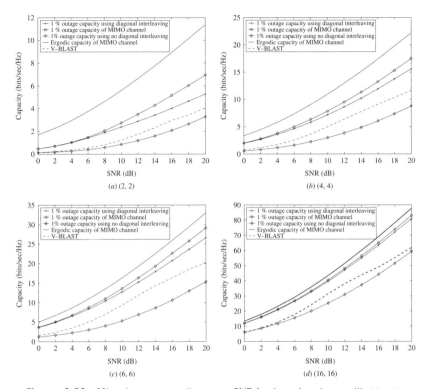

Figure 3.22 1% outage capacity versus SNR for (n_t, n_r) systems with $M = N$.

The figures also show the significant capacity gain achieved by the scheme with DSTI compared to V-BLAST. In Figure 3.23, we illustrate the per-layer rates separately for $n_t = n_r = 2$ and 4. The larger n_t and n_r are, the better the performances of the scheme with DSTI. It is worth mentioning that if we increase n_t for fixed n_r, then the average MI (ergodic capacity) approaches a limit and its variance decreases. Moreover, if we increase n_r for fixed n_t, then the average MI increases logarithmically and its variance decreases; see [72]. Therefore, if the number of antennas increases, the variance decreases and hence the density of the MI gets more concentrated around its mean (i.e, asymptotically, the outage and ergodic capacities coincide).

3.9 SUMMARY AND DISCUSSION

In this chapter, four popular MIMO systems were described. All of these methods have advantageous and drawbacks:

- The D-BLAST architecture is a theoretical superstructure capable of attaining the Shannon capacity with 1-D codec technology but unrealizable in practice.

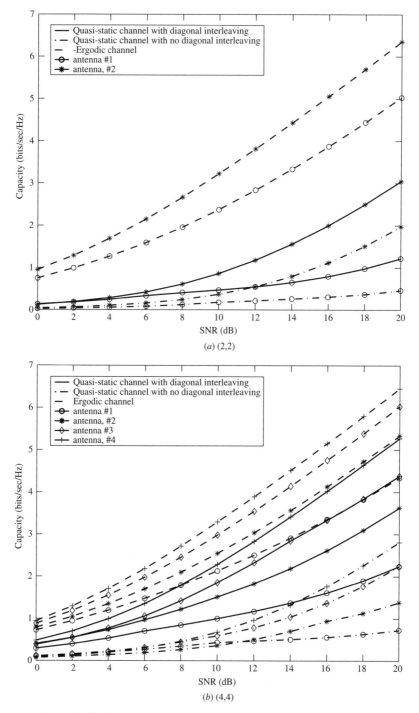

Figure 3.23 1% outage per-layer rates versus SNR for (M,M) systems.

- The V-BLAST architecture is physically implementable but is susceptible to error propagation when any of its antennas enters a deep fade.
- In SD-BLAST, signal constituents are arranged in space-time as helically wrapped plies. They are so arranged to enable 1-D signal processing, so the receiver can substantially mute inevitable mutual interference from the n_t simultaneous transmissions. Using uncoordinated 1-D codecs, SD-BLAST avoids waste of the space-time resource. An information theoretic analysis showed that this communication architecture can often approach the outage capacity performance, particularly in the cases where n_t is significantly greater than n_r, i.e., where the receiver has few antennas compared to the base(s).
- The MLST codes were designed with the objective of using moderate to large numbers of transmit and receive antennas with a decoding complexity less than that of V-BLAST. In contrast to the other three BLAST architectures, the MLST architectures assign unequal code rates to the layers and code rates are decided on the basis of decoding order. In particular, the proposed decision feedback interference cancellation receivers can achieve nearly optimal performances for $n_r = n_t \geq 16$.

For a small sampling of (n_t, n_r) examples involving matrix Rayleigh channels, we quantified the extent to which one can expect to approach the capacity limits with SD-BLAST using Monte Carlo simulations. Moreover, for SD-BLAST, under the assumption of equal energy per stratum, we quantified how much stratification was needed to support a particular capacity level at the 10% outage level for SNRs ranging from -6 to $+24$ dB. Although in theory the channel-hardening approximation improves for large n_t and n_r, the outage capacity performance of SD-BLAST shows significant channel hardening even for small n_t and n_r.

The computational effort associated with the onion-peeling receiver of SD-BLAST is similar to that of the decoders of the D-BLAST architectures (interference avoidance and cancellation) but scales up with the number of plies L. The computational complexity associated with threaded-BLAST also scales up with the number of iterations. V-BLAST has the computational issue of finding the optimal ordering for the sequential detection. In this chapter, we did not consider the details of 1-D code designs for SD-BLAST architectures. This is an interesting and important topic. Channel code designs for D-BLAST have been proposed in [89], where trellis codes have been developed to have a distance structure that is matched with periodic SNR variation of the channel created by the diagonal coding introduced in D-BLAST. V-BLAST and threaded-BLAST can work with codes designed for a standard AWGN environment. To minimize the effect of decision errors in V-BLAST and threaded-BLAST and to improve the detection and decoding gain, iterative detection and decoding (IDD) are incorporated in many references with significant improvement in bit error rate performance [81].

The design of channel codes for SD-BLAST faces the following challenges:

- The design of groups of multirate codes, which is complicated because each ply in SD-BLAST faces a different SNIR level according to the proposed receiver signal processing.
- The design of channel codes for systems with periodically varying SNRs, which is a challenging problem partially addresses by Matache et al. in [89].

Reference [89] discusses coding for systems with periodically varying SNRs; there it is stressed that the signaling alphabet must be sufficiently rich to take advantage of the better SNRs. For SD-BLAST, with more and more plies, even a code using a binary constellation becomes sufficiently rich as L increases.

Finally, we described a class of multirate diagonal space-time interleaved codes and successive decoding and interference cancellation receivers for the quasi-static Rayleigh fading channels. These codes were designed to use moderate to large numbers of transmit and receive antennas with a decoding complexity less than that of V-BLAST. In contrast to the other three BLAST architectures, the MLST architectures use unequal code rates to the layers and code rates are decided based on the decoding order. Using the probability density function of the per-layer, we analyzed the outage capacity performances of the proposed scheme. In particular, the proposed decision FB interference cancellation receivers can achieve nearly optimal performances for $n_r = n_t \geq 16$. Moreover, for $n_r = n_t \geq 4$, the performance loss of the proposed scheme compared to the optimum performance is less than that reported for V-BLAST and similar schemes.

APPENDIX: OPTIMALITY OF D-BLAST

Consider an (M, N) system. The goal is to prove

$$\log_2 \det \left(\mathbf{I}_N + \frac{\rho}{M} \mathbf{H} \mathbf{H}^H \right) = \sum_{k=1}^{M} \log_2 [1 + \Upsilon_k] \tag{3.75}$$

The above proof may be simplified as

$$\det \left(\mathbf{I}_N + \frac{\rho}{M} \mathbf{H} \mathbf{H}^H \right) = \prod_{k=1}^{M} [1 + \Upsilon_k] \tag{3.76}$$

Define

$$\Omega_M = \left(\mathbf{I}_N + \frac{\rho}{M} \mathbf{H} \mathbf{H}^H \right) = \frac{\rho}{M} \sum_{i=1}^{M} \mathbf{h}_i \mathbf{h}_i^H + \mathbf{I}_N \tag{3.77}$$

where $\mathbf{h}_i$ is ith column of matrix $\mathbf{H}$.

For single transmit and N receive antennas (receiver diversity only), it can be shown that

$$\det\left(1 + \frac{\rho}{M}\mathbf{h}_1^H \mathbf{h}_1\right) = 1 + \Upsilon_1 \tag{3.78}$$

Let us assume that (3.76) is true for $m - 1$ transmit antennas. Then we can write

$$\det(\Omega_{m-1}) = \prod_{k=1}^{m-1}[1 + \Upsilon_k] \tag{3.79}$$

From (3.77), we have

$$\Omega_m = \Omega_{m-1} + \frac{\rho}{M}\mathbf{h}_m\mathbf{h}_m^H \tag{3.80}$$

$$\mathbf{P} = \mathbf{Q} + \frac{\rho}{M}\mathbf{h}_m\mathbf{h}_m^H \tag{3.81}$$

where we have defined: $\mathbf{P} = \Omega_m$ and $\mathbf{Q} = \Omega_{m-1}$. Using the matrix inversion lemma given in [7] and [94], we can show that

$$[\mathbf{Q} + \frac{\rho}{M}\mathbf{h}_m\mathbf{h}_m^H]^{-1}\mathbf{h}_m = \frac{\mathbf{Q}^{-1}\mathbf{h}_m}{1 + \frac{\rho}{M}\mathbf{h}_m^H\mathbf{Q}^{-1}\mathbf{h}_m}$$

$$\mathbf{P}^{-1}\mathbf{h}_m = \frac{\mathbf{Q}^{-1}\mathbf{h}_m}{1 + \Upsilon_m} \tag{3.82}$$

The matrix inverse $\mathbf{P}^{-1}$ is defined by [57]:

$$\mathbf{P}^{-1} = \frac{\mathbf{P}_{adj}}{\det(\mathbf{P})} \tag{3.83}$$

where $\mathbf{P}_{adj}$ is the adjoint matrix of matrix $\mathbf{P}$; thus, we can rewrite (3.82) as

$$\frac{\mathbf{P}_{adj}}{\det(\mathbf{P})}\mathbf{h}_m = \frac{\mathbf{Q}_{adj}}{\det(\mathbf{Q})}\frac{\mathbf{h}_m}{1 + \Upsilon_m} \tag{3.84}$$

$$= \mathbf{Q}_{adj}\frac{\mathbf{h}_m}{\prod_{i=1}^{m}(1 + \Upsilon_i)} \tag{3.85}$$

where (3.85) is obtained by replacing $\det\mathbf{Q}$ using (3.79). Now, to prove (3.76), we have to show that

$$\mathbf{P}_{adj}\mathbf{h}_m = \mathbf{Q}_{adj}\mathbf{h}_m \tag{3.86}$$

Expanding (3.81), we have

$$\mathbf{P} = \mathbf{Q} + \frac{\rho}{M}\mathbf{h}_m\mathbf{h}_m^H$$

$$= \left[\mathbf{q}_1 + \frac{\rho}{M}\mathbf{h}_m H_{m1}^*, \mathbf{q}_2 + \frac{\rho}{M}\mathbf{h}_m H_{m2}^*, \ldots, \mathbf{q}_N + \frac{\rho}{M}\mathbf{h}_m H_{mN}^*\right] \tag{3.87}$$

where $\mathbf{q}_i$, $\forall i$ is the ith column of $\mathbf{Q}$ and H_{ij} is the ijth element of matrix $\mathbf{H}$. We can prove (3.86) by showing that the jth element of $\mathbf{P}_{adj}\mathbf{h}_m$ is equal to the jth element of $\mathbf{Q}_{adj}\mathbf{h}_m$ for $j = 1, \ldots, m$. The jth element of $\mathbf{P}_{adj}\mathbf{h}_m$ is given by $\det(\mathbf{P}_j)$, where $\mathbf{P}_j$ is obtained by replacing the jth column of matrix $\mathbf{P}$ by $\mathbf{h}_m$. Similarly, the jth element of $\mathbf{Q}_{adj}\mathbf{h}_m$ is given by $\det(\mathbf{Q}_j)$. Accordingly, by replacing the jth column of (3.88) by $\mathbf{h}_m$, we get

$$\det(\mathbf{P}_j) = \det\left[\mathbf{q}_1 + \frac{\rho}{M}\mathbf{h}_m H_{m1}^*, \ldots, \mathbf{h}_m, \ldots, \mathbf{q}_N + \frac{\rho}{M}\mathbf{h}_m H_{mN}^*\right] \tag{3.88}$$

Using the fact that a determinant does not change if one column multiplied by a constant is added to another column, we can rewrite (3.88) as

$$\det(\mathbf{P}_j) = \det[\mathbf{q}_1, \mathbf{q}_2, \ldots, \mathbf{h}_m, \ldots, \mathbf{q}_N] \tag{3.89}$$

$$= \det(\mathbf{Q}_j) \tag{3.90}$$

Similarly, we can show that $\det(\mathbf{P}_j) = \det(\mathbf{Q}_j)$ for $i = 1, 2, \ldots, N, i \neq j$. This completes the proof of (3.75).

4

SPACE-TIME TURBO CODES AND TURBO DECODING PRINCIPLES

4.1 INTRODUCTION

In 1993, Berrou et al. developed the revolutionary iterative turbo receiver for decoding concatenated codes [16].[1] Three properties are the hallmark of turbo codes:

- The design of an encoder that approximates a random-like encoded message.
- The error performance of the turbo decoder, in which performance improves with the number of iterations of the decoding algorithm.
- The ability of the turbo decoder to approach the Shannon limit of channel capacity in a computationally feasible manner.

What is even more profound is the fact that the turbo processing principle, or turbo principle for short, has been successfully applied not only to channel decoding, but also to channel equalization, coded modulation, multiuser detection, and joint source and channel decoding [62]. In this chapter, we describe turbo codes and iterative decoding principles. We also present an application of the turbo code and the turbo processing principle to wireless communications when using multiple antennas at both the transmitting and receiving ends of the wireless channel. This

---

[1]An important contribution of the work that preceded the work of Berrou et al. on turbo codes is that of Lodge et al. on iterative decoders for 2-D product-like codes [84].

Space-Time Layered Information Processing for Wireless Communications,
By Mathini Sellathurai and Simon Haykin
Copyright © 2009 John Wiley & Sons, Inc.

antenna structure is called turbo space-time codes or space-time turbo codes. In particular, we describe

- serial and parallel concatenated turbo codes and their iterative decoders,
- soft-in/soft-out modules, which are exemplified by the BCJR algorithm that performs maximum *a posteriori* estimation on a bit-by-bit basis in the decoding of turbo codes and their lower-complexity and numerically less sensitive approximations,
- the extraction of extrinsic information, which is believed to be behind the success of turbo principles, and
- space-time turbo codes.

4.2 TURBO CODES

The turbo encoder consists of two simple convolutional encoders. A block of bits is encoded by two simple recursive convolutional codes, each with a relatively small number of states. The input to the second encoder is an interleaved (i.e., pseudorandomized) version of the bits output from the first encoder.

The turbo code is thus the combination of uncoded bits and parity bits generated by the two encoders. An innovative feature of turbo codes is the use of a random interleaver, which permutes the original block of bits before application to the second encoder. The high error-correcting power of turbo codes originates from random-like coding achieved by random interleaving in conjunction with concatenated coding and iterative decoding using uncorrelated extrinsic information. The random-like structure of turbo codes has been shown to produce outstanding performance by providing small error rates at data rates close to the theoretical channel capacity. The structure and complexity of turbo encoder design are restricted by system parameters such as *decoding delay* and *coding gain*.

4.2.1 Parallel Concatenated Turbo Codes

In parallel concatenated turbo codes, two systematic convolutional codes with rates R_1 and R_2 are concatenated in parallel in the form shown in Figure 4.1. Moreover, the block of L bits is fed directly to the first encoder. For the second encoder, the same block of data is permuted using a pseudoblock interleaver, represented by the symbol Π. These two encoders are not necessarily identical, but for best decoding performance, $R_1 \leq R_2$.

The block diagram of the iterative decoder, depicted in Figure 4.2, is made up of two elementary soft-in/soft-out decoder modules denoted by "SISO," one for each encoder, an interleaver and a de-interleaver that performs the inverse permutation with respect to the interleaver.

At the core of the decoding algorithm is the SISO module. A SISO module is a four-port device that accepts the probability distributions or the corresponding likelihood ratios of the information and noisy encoded symbols as inputs and provides an update of these probability distributions based upon the code constraint as

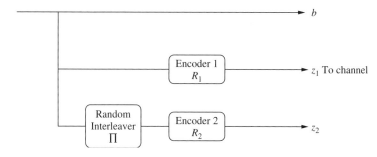

Figure 4.1 Turbo codes.

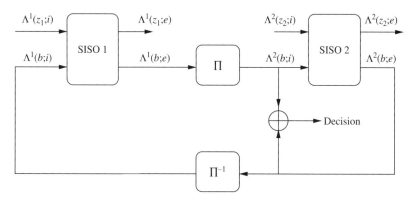

Figure 4.2 Turbo decoder.

outputs [14]. The symbols $\Lambda(:, i)$ and $\Lambda(:, e)$ at the input and output ports of the SISO module refer to the *intrinsic* and *extrinsic* information computed in terms of log-likelihood ratios (LLR).

Definition 1 *The LLR of a binary random variable U in GF(2) with the elements* $\{+1, -1\}$ *is defined as*

$$\Lambda(u) = \log \frac{P(u = +1)}{P(u = -1)} \tag{4.1}$$

The sign and the magnitude of $\Lambda(u)$ *correspond to a hard decision and its reliability.*

Notations used in LLR are summarized here. The first argument refers to the information symbols (u) or the parity bits (z_1 or z_2). The second argument refers to intrinsic (i), extrinsic (e), or *a posteriori* (p) information. Finally, the superscripts 1 and 2 refer to the SISO decoders 1 and 2, respectively.

The turbo principle makes use of the concepts of intrinsic and extrinsic information to increase the independence of inputs from one processing stage to the next. A SISO module accepts intrinsic or *a priori* values and provides the *a posteriori* or extrinsic information. Before describing the operation of the turbo decoder in detail, we define intrinsic and extrinsic information.

Definition 2 *Intrinsic information refers to the soft information inherent in a bit received over the channel. Typically, it is the sample* a priori *values prior to decoding, i.e., unconstrained probability, as shown by*

$$\Lambda(u; i) = \log \frac{P\{u = +1\}}{P\{u = -1\}} \qquad (4.2)$$

Definition 3 A posteriori *information is the information provided about a bit u from the received bits according to the code constraint*

$$\Lambda(u; p) = \log \frac{P\{u = +1|decoding\}}{P\{u = -1|decoding\}} \qquad (4.3)$$

where P{u|decoding} is the probability of information bit u computed by the decoder using the knowledge of channel code constraint.

Definition 4 *The extrinsic information is the information provided about a bit u from the other received bits according to the constraint imposed by the FEC code. It is formally defined as the difference between the* a posteriori *information and the intrinsic information fed back to the input of the decoding stage, as shown by*

$$\Lambda(u; e) = \Lambda(u; p) - \Lambda(u; i) \qquad (4.4)$$

The exchange of extrinsic informations between the decoding stages of the receiver assists in the slow but steady convergence of the iterative decoding process. Specifically, with increasing iterations, the extrinsic values achieve better and better confidence levels. At convergence, the extrinsic informations will be directed toward the sign of the information bits and with high reliability. (See [140].)

The iterative decoder illustrated in Figure 4.2 depicts the flow of a message passing between the SISO modules. The SISO module operations can be explained using the following three steps:

1. The first SISO module generates soft estimates of the systematic bits $b(l)$, $l = 1, 2, \ldots, L$:
   - First, estimate the *a posteriori information*

$$\Lambda^1(b(l); p) = \log \frac{P\{b(l) = +1|\Lambda^1(\mathbf{z_1}; i), \Lambda^1(\mathbf{b}; i), decoding\}}{P\{b(l) = -1|\Lambda^1(\mathbf{z_1}; i), \Lambda^1(\mathbf{b}; i), decoding\}},$$
$$l = 1, 2, \ldots L \qquad (4.5)$$

During the first iteration $\Lambda^1(\mathbf{z_1}; i)$ and $\Lambda^1(\mathbf{b}; i)$ are initialized, with the soft outputs consisting of the LLRs of the information symbols $\mathbf{b}$ and the first set of parity check bits $\mathbf{z}_1$ for the received channel signal [140].

- Compute the extrinsic information

$$\Lambda^1(\mathbf{b}; e) = \Lambda^1(\mathbf{b}; p) - \Lambda^1(\mathbf{b}; i) \qquad (4.6)$$

- $\Lambda^1(\mathbf{b}; e)$ is the extrinsic information about the set of message bits $\mathbf{b}$ derived from the first decoding stage and fed to the second decoding stage as the intrinsic information. Before proceeding to the second decoding stage, the extrinsic information is reordered to compensate for the pseudorandom interleaving introduced in the turbo encoder:

$$\Lambda^2(\mathbf{b}; i) = \Pi\{\Lambda^1(\mathbf{b}; e)\} \qquad (4.7)$$

2. Similarly, the second SISO module associated with the second encoder provides a further refined estimate of the systematic bit $b(l)$ using the soft outputs consisting of the LLRs of the interleaved information symbols $\mathbf{b}$ and the second set of parity check bits $\mathbf{z}_2$ for the received channel signal [141]. The output of the second stage provides intrinsic information to the first stage after reordering to compensate for the random interleaving operation, i.e.,

$$\Lambda^1(\mathbf{b}; i) = \Pi^{-1}\{\Lambda^2(\mathbf{b}; e)\} \qquad (4.8)$$

Steps 1 and 2 are repeated until the algorithm converges.[2]

3. The estimate of the message bits $\mathbf{b}$ is obtained by hard limiting the LLR $\Lambda^2(\mathbf{b}; p)$ at the output of the second stage:

$$\hat{\mathbf{b}} = \text{sgn}\{\Lambda^2(\mathbf{b}; p)\} \qquad (4.9)$$

Note the following:
- The estimate $\hat{\mathbf{b}}$ approaches the global maximum *a posteriori* (MAP) solution as the number of iterations approaches infinity, provided that the bit probabilities remain independent between iterations of the decoding process. Here, we mean global MAP by decoding a single Markov process[3] (trellis) modeled for the turbo code, which includes the effects of the interleaver.
- The fundamental principle for feeding back the extrinsic information from one decoder to another is to never feed back decoder information that stems from itself. The feedback of extrinsic information prevents the enhancement of highly correlated input and output corruptions. Typically, for cases encountered in practice, the feedback of extrinsic information maintains the statistical independence between stages.

---

[2]In practice, we may do one of two things: Either steps 1 and 2 are repeated a fixed number of times or appropriate stopping criteria are used [15].
[3]A Markov process is a process that can be in one of several states and can pass from one state to another at each time step according to fixed probabilities. If a Markov process is in state i, there is a fixed probability, p_{ij}, of its going into state j at the next time step; p_{ij} is called a transition probability and can be illustrated by means of a state transition diagram, which is a diagram showing all the states and transition probabilities.

- Finally, the separation of extrinsic information from the estimate of *a posteriori* probabilities holds only if the inputs to the decoders are independent. If the channel has memory, the independence assumption is not valid; therefore, interleaving between the decoder stages is important. For better performance, either an outer interleaver or irregular turbo codes may be used. In particular, in a standard turbo code, the interleaver maps each systematic bit to a unique input bit of convolutional encoder 2. In contrast, irregular turbo codes (first introduced by Frey and MacKay [50]) use a special design of interleaver that maps some systematic bits to multiple input bits of the convolutional encoder. The irregular turbo codes are typically described by using a degree profile $(f_d, d \in 1, 2, \ldots, D)$, where f_d is the fraction of the bits that have degree d. The parameter D is the maximum degree. Each information bit with degree d is repeated d times before being interleaved. For example, each bit of 10% of the systematic bits may be mapped to eight inputs of the convolutional encoder instead of a single one. The decoder of the irregular turbo codes has a decoder structure similar to that of a regular turbo codes ([52], [68]).

4.2.2 Serial Concatenated Turbo Codes

An alternative to parallel concatenated turbo codes is serially concatenated convolutional codes formed by concatenating two systematic convolutional codes in the form shown in Figure 4.3. The incoming block of information bits is encoded by the first encoder. Then the interleaver permutes the output codewords of the outer code before passing them on to the inner code [14]:

$$\mathbf{b}^i = \Pi(\mathbf{c}^o) \tag{4.10}$$

These codes may be viewed as a new class of turbo codes since they can be decoded using iterative decoders.

The decoder for the serially concatenated codes is a concatenation of inner and outer SISO modules, as shown in Figure 4.4. The figure clearly depicts the flow of the message passing between the inner and outer SISO modules. The decoding process involves the following computations:

1. The inner SISO module generates a soft estimate of the information bits $\mathbf{b}^i$ conditioned on the inner code constraint:

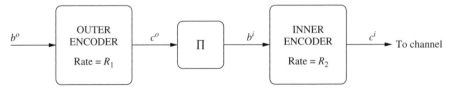

Figure 4.3 Serially concatenated convolutional codes.

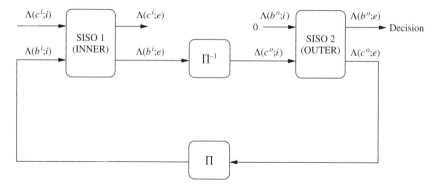

Figure 4.4 Iterative decoder for serially concatenated codes.

- Estimation of the *a posteriori* information:

$$\Lambda(b^i(l); p) = \log \frac{P\{b^i(l) = +1|\Lambda(\mathbf{c}^i; i), \Lambda(\mathbf{b}^i; i), \text{decoding}\}}{P\{b^i(l) = -1|\Lambda(\mathbf{c}^i; i), \Lambda(\mathbf{b}^i; i), \text{decoding}\}},$$

$$l = 1, 2, \ldots L \qquad (4.11)$$

During the first iteration, $\Lambda(\mathbf{c}^i; i)$ is initialized, with the soft outputs consisting of the LLRs of symbols received from the channel. The second input, $\Lambda(\mathbf{b}, i)$, is initialized to zero since no *a priori* information is available on the input symbols $\mathbf{b}$.

- Computation of the extrinsic information

$$\Lambda(\mathbf{b}^i; e) = \Lambda(\mathbf{b}^i; p) - \Lambda(\mathbf{b}^i; i) \qquad (4.12)$$

- $\Lambda(\mathbf{b}^i; e)$ is the extrinsic information about the set of message bits $\mathbf{b}^i$ of the inner encoder and fed back to the outer decoder as the intrinsic information of its coded bits. Before the application of the outer decoder, the extrinsic information is reordered to compensate for the pseudorandom interleaving introduced in the turbo encoder:

$$\Lambda(\mathbf{c}^o; i) = \Pi^{-1}\{\Lambda(\mathbf{b}^i; e)\} \qquad (4.13)$$

2. The outer SISO, in turn, processes the LLR's $\Lambda(\mathbf{c}^o; i)$ and computes the LLRs of both code and information symbols based on the outer code constraints.
   - The *a posteriori* information for information and code symbols is defined by

$$\Lambda(b^o(l); p) = \log \frac{P\{b^o(l) = +1|\Lambda(\mathbf{c}^o; i), \Lambda(\mathbf{b}^o; i), \text{decoding}\}}{P\{b^o(l) = -1|\Lambda(\mathbf{c}^o; i), \Lambda(\mathbf{b}^o; i), \text{decoding}\}},$$

$$l = 1, 2, \ldots L \qquad (4.14)$$

$$\Lambda(c^o(l); p) = \log \frac{P\{c^o(l) = +1 | \Lambda(\mathbf{c}^o; i), \Lambda(\mathbf{b}^o; i), \text{decoding}\}}{P\{c^o(l) = -1 | \Lambda(\mathbf{c}^o; i), \Lambda(\mathbf{b}^o; i), \text{decoding}\}},$$
$$(4.15)$$
$$l = 1, 2, \ldots L$$

The input $\Lambda(\mathbf{b}^o, i)$ is always initialized to zero, assuming equally likely source information symbols.

- Extrinsic information of both the information and the code symbols:

$$\Lambda(\mathbf{b}^o; e) = \Lambda(\mathbf{b}^o; p) - \Lambda(\mathbf{b}^o; i) \qquad (4.16)$$

$$\Lambda(\mathbf{c}^o; e) = \Lambda(\mathbf{c}^o; p) - \Lambda(\mathbf{c}^o; i) \qquad (4.17)$$

The output of the outer decoder provides intrinsic information to the inner decoder after reordering to compensate for the random interleaving:

$$\Lambda(\mathbf{b}^i; i) = \Pi\{\Lambda(\mathbf{c}^o; e)\} \qquad (4.18)$$

Steps 1 and 2 are repeated until the algorithm converges.

3. An estimate of the message bits $\mathbf{b}^o$ is obtained by hard limiting the LLR $\Lambda(\mathbf{b}^o; p)$ at the output of the outer SISO:

$$\hat{\mathbf{b}}^o = \text{sgn}(\Lambda(\mathbf{b}^o; p)) \qquad (4.19)$$

Note the following:
- In contrast to the turbo decoder used for traditional turbo codes, which updates only the LLR of systematic symbols, the iterative decoder associated with the serially concatenated turbo codes updates the LLR of both information and code symbols based on the code constraint.
- The interleaver gain of serially concatenated codes, defined as the factor that decreases the bit error probability as a function of interleaver size, can be made significantly higher than that of traditional turbo codes [14].

4.2.3 SISO Decoders

This section introduces a key feature in the iterative decoders; the MAP-based SISO/module and, in particular, an implementation technique known as a generalized BCJR algorithm [14].

Historical Remarks In 1974, the pioneering work of Bahl, Cocke, Jelinek, and Raviv established the symbol-by-symbol MAP algorithm as an alternative to the Viterbi algorithm for decoding convolutional codes. The MAP algorithm performs forward and backward recursions which, in [22], were used for canceling intersymbol interference (ISI). Earlier, in 1970, a similar technique was used to perform forward recursion only for canceling ISI, but typically it had higher memory and complexity requirements [1]. The MAP algorithm requires the whole sequence for decoding and thus can be used for decoding only in block mode. Despite a long delay, the memory requirement and computational complexity grow only linearly with the sequence length. However, the algorithm requires a forward recursion; hence, its memory and computational complexity grow exponentially with decoding delay.

4.2.4 Generalized BCJR Algorithm

The SISO modules used in the turbo processing uses the BCJR algorithm, so called in honor of its four inventors: Bahl, Cocke, Jelinek, and Raviv [9]. The BCJR algorithm is fundamentally different from the Viterbi algorithm:

- The Viterbi algorithm is a maximum likelihood sequence estimator in that it maximizes the likelihood function for the whole sequence (usually a code-word). Thus, it minimizes the sequence or codeword error rate of a communication system.
- The BCJR algorithm is a MAP decoder in that it minimizes the bit errors by estimating the *a posteriori* probabilities of the individual bits in a codeword. The average bit error rate of the BCJR algorithm can be slightly better than that of the Viterbi algorithm, but it can never be worse.

The formulation of the BCJR algorithm relies on the assumption that the channel is memoryless and that the channel encoder and the channel coding can be completely described by a Markov process. This means that if a code can be represented as a trellis, then the present state of the trellis depends only on the past state and the input bit [68]. The trellis of a block code or convolutional code can always be made to start and end in the zero state by adding zero bits to the information sequence appropriately. Moreover, the dynamics of the time-invariant trellis are completely specified by a trellis section [14]. A trellis section is depicted in Figure 4.5.

Let $\mathbf{b} = \{b(t)\}_{t=1}^{\tau}$ be the input to a trellis encoder drawn from the alphabet $\mathcal{B} = \{b_1, \ldots, b_{N_I}\}$, let $\mathbf{c} = \{c(t)\}_{t=1}^{\tau}$ be the sequence of output or code from the trellis encoder drawn from the alphabet $\mathbb{C} = \{c_1, \ldots, c_{N_O}\}$, and let $\mathbf{y} = \{y(t)\}_{t=1}^{\tau}$ be the corresponding output observed at the receiver.

Definition 5 *A trellis section can be characterized by a set of N_s states $\mathcal{S} = \{s_1, \ldots, s_{N_s}\}$ and a set of N_I edges $\mathcal{E} = \{e_1, \ldots, e_{N_s \times N_I}\}$ between the states of the trellis at time t and time $t + 1$. An edge can be identified by the information symbol $b(e)$ and the code symbol $c(e)$ associated with the edge, and the starting (S) and ending (E) states of the edge are denoted as $s^S(e)$ and $s^E(e)$, respectively.*

We consider a new generalized form of the BCJR algorithm which is suitable for any component or block codes represented by a trellis [14]. The generalized BCJR algorithm can cope with codes having a trellis with parallel edges. Note that the original BCJR algorithm could not cope with a trellis having parallel edges.

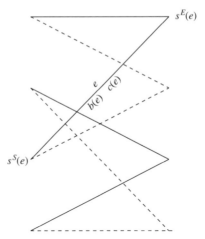

Figure 4.5 Trellis section between times t and $t+1$.

Consider a trellis section between time t and time $t+1$. The BCJR algorithm first computes the *a posteriori* probabilities of the states and transition of the trellis given by the conditional probability $P\left(s^S(e) \to s^E(e)|\mathbf{y}\right)$ of each valid state transition given the noisy channel observations $\mathbf{y}$:

$$P_t\left(s^S(e) \to s^E(e))|\mathbf{y}\right) = \frac{P_t\left(s^S(e) \to s^E(e), \mathbf{y}\right)}{P_t(\mathbf{y})} \qquad (4.20)$$

This joint probability $P_t\left(s^S(e) \to s^E(e), \mathbf{y}\right)$ can be partitioned using the properties of a Markov process. By definition, for a Markov process, the state distribution at time t, given the past, is independent of the state distribution at time t, given the future [95], [153]:

$$P_t\left(s^S(e) \to s^E(e), \mathbf{y}\right) = \alpha_t\left(s^S(e)\right)\gamma_t\left(s^S(e) \to s^E(e)\right)\beta_{t+1}\left(s^E(e)\right) \qquad (4.21)$$

where

$$\alpha_t\left(s^S(e)\right) = P_t\left(s^S(e), (y_0, \dots y_{t-1})\right) \qquad (4.22)$$

$$\gamma_t\left(s^S(e) \to s^E(e)\right) = P_{t+1}\left(s^E(e), y_t|s^S(e)\right) \qquad (4.23)$$

$$\beta_{t+1}\left(s^E(e)\right) = P_{t+1}\left(s^E(e)|(y_{t+1}, \dots y_\tau)\right) \qquad (4.24)$$

Equations 4.22 and 4.23 are the estimates of the state probabilities at time t based on past and future observations, respectively, and the state transition probability at time t can be expressed as

$$\gamma_t\left(s^S(e) \to s^E(e)\right) = P_{t+1}\left(s^E(e)|s^S(e)\right) P_{t+1}\left(y_t|s^S(e) \to s^E(e)\right) \qquad (4.25)$$

The first probabilistic factor on the right-hand side of (4.25) follows from the fact that, given $s^S(e)$, the probability of going to state $s^E(e)$ depends only on the probability of the information bit at time t. The second probabilistic factor on the right-hand side of (4.25) follows from the fact that, if we know $s^S(e)$ and $s^E(e)$, then since the code is deterministic, the code symbol is known. Thus, (4.25) reduces to

$$\gamma_t\left(s^S(e) \to s^E(e)\right) = P_{t+1}(b(e))P_{t+1}(y_t|c(e)) \tag{4.26}$$

Forward and Backward Recursions The forward and backward estimates can be determined recursively as follows, respectively:

$$
\begin{aligned}
\alpha_t(s) &= \sum_{e:s^E(e)=s} \alpha_{t-1}\left[s^S(e)\right]\gamma_{t-1}\left[s^S(e) \to s^E(e)\right] \\
&= \sum_{e:s^E(e)=s} \alpha_{t-1}\left[s^S(e)\right]P_t[b(e), i]P_t[c(e); i], t = 1, 2, \ldots, \tau
\end{aligned}
\tag{4.27}
$$

$$
\begin{aligned}
\beta_t(s) &= \sum_{e:s^S(e)=s} \beta_{t+1}\left[s^E(e)\right]\gamma_t\left[s^S(e) \to s^E(e)\right] \\
&= \sum_{e:s^S(e)=s} \beta_{t+1}\left[s^E(e)\right]P_{t+1}[b(e); i]P_{t+1}[c(e); i], t = \tau - 1, \tau - 2, \ldots 0
\end{aligned}
\tag{4.28}
$$

$\alpha_0(s)$ and $\beta_\tau(s)$ are intialized according to the trellis structure. For a trellis starting and ending in the all-zero state, we have

$$\alpha_0(s) = \begin{cases} 1, & s = S_0 \\ 0 & \text{otherwise} \end{cases} \tag{4.29}$$

$$\beta_\tau(s) = \begin{cases} 1, & s = S_\tau \\ 0 & \text{otherwise} \end{cases} \tag{4.30}$$

A Posteriori Probabilities At time t, the output *a posteriori* probabilities (APPs) of the information and code bits are found according to the following two formulas, respectively:

$$P_t(c; p) = \sum_{e:c(e)=c} \alpha_{t-1}\left[s^S(e)\right]P_t[b(e); i]P_t[c(e); i]\beta_t\left[s^E(e)\right] \tag{4.31}$$

$$P_t(b; p) = \sum_{e:b(e)=b} \alpha_{t-1}\left[s^S(e)\right]P_t[b(e); i]P_t[c(e); i]\beta_t\left[s^E(e)\right] \tag{4.32}$$

Extrinsic Information From the APP defined in (4.31) and (4.32), the extrinsic information can be obtained by extracting $P_t(c(e), i)$ and $P_t(b(e), i)$, respectively.

Note that these two probabilities do not depend on e; they can therefore be extracted.

$$P_t[c; e] = a_c \frac{P_t(c; p)}{P_t(c; i)} \tag{4.33}$$

$$P_t(b; e) = a_b \frac{P_t(b; p)}{P_t(b; i)} \tag{4.34}$$

The normalization factor a_b in (4.34) is obtained by ensuring that the sum of the probabilities of information bits is equal to unity:

$$a_b \leftarrow \sum_b P_t(b; e) = 1 \tag{4.35}$$

Similarly, the normalization factor a_c in (4.33) is defined by writing

$$a_c \leftarrow \sum_c P_t(c; e) = 1 \tag{4.36}$$

Correspondingly, the extrinsic information takes the following forms:

$$P_t(c; e) = a_c \sum_{e:c(e)=c} \alpha_{t-1}\big[s^S(e)\big] P_t[b(e); i] \beta_t\big[s^E(e)\big] \tag{4.37}$$

$$P_t(b; e) = a_u \sum_{e:u(e)=u} \alpha_{t-1}\big[s^S(e)\big] P_t[c(e); i] \beta_t\big[s^E(e)\big] \tag{4.38}$$

4.2.5 The MAP Algorithm in the Log Domain (LOG-MAP Algorithm)

The MAP algorithm has high computational complexity: $2 \times N_s \times N_I$ multiplications and additions per bit. This can be reduced by performing the algorithm in the log domain, where multiplication becomes summation [123]. First, we define the following terms:

1. The transition probability:

$$
\begin{aligned}
C_{t-1}\big[s^S(e) \rightarrow s^E(e)\big] &= \log \gamma_{t-1}\big[s^S(e) \rightarrow s^E(e)\big] \\
&= \log P_t[b(e), i] + \log P_t[c(e); i] \qquad (4.39) \\
&= \pi_t[b(e), i] + \pi_t[c(e); i]
\end{aligned}
$$

2. The forward estimator for $t = 1, 2, \ldots, \tau$:

$$A_t(s) = \log \alpha_t\left(s^S(e)\right)$$

$$= \log\left[\sum_{e:s^E(e)=s} \left\{\alpha_{t-1}\left[s^S(e)\right] \cdot P_t[b(e); i] \cdot P_t[c(e); i]\right\}\right]$$

$$= \log\left[\sum_{e:s^E(e)=s} \exp\left\{A_{t-1}\left[s^S(e)\right] + \pi_t[b(e); i] + \pi_t[c(e); i]\right\}\right] (4.40)$$

$$= \max_{e:s^E(e)=s}^{*} \left\{A_{t-1}\left[s^S(e)\right] + \pi_t[b(e); i] + \pi_t[c(e); i]\right\}$$

with initial values

$$A_0(s) = \begin{cases} 0, & s = S_0 \\ -\infty, & \text{otherwise} \end{cases} \qquad (4.41)$$

3. The backward estimator for $t = \tau - 1, \tau - 2, \ldots, 0$:

$$B_t(s) = \log \beta_t(s)$$

$$= \log\left[\sum_{e:s^S(e)=s} \left\{\beta_{t+1}\left[s^E(e)\right] \cdot P_{t+1}[b(e); i] \cdot P_{t+1}[c(e); i]\right\}\right]$$

$$= \log\left[\sum_{e:s^S(e)=s} \exp\left\{B_{t+1}\left[s^E(e)\right] + \pi_{t+1}[b(e); i] + \pi_{t+1}[c(e); i]\right\}\right] (4.42)$$

$$= \max_{e:s^S(e)=s}^{*} \left\{B_{t+1}\left[s^E(e)\right] + \pi_{t+1}[b(e); i] + \pi_{t+1}[c(e); i]\right\}$$

with initial values

$$B_\tau(s) = \begin{cases} 0 & s = S_\tau \\ -\infty & \text{otherwise} \end{cases} \qquad (4.43)$$

Once forward and backward estimates are computed, the output extrinsic information is found by using the following two formulas:

$$\pi_t(c; e) = \log\left[\sum_{e:c(e)=c} \left\{\alpha_{t-1}\left[s^S(e)\right] \cdot P_t[b(e); i] \cdot \beta_{t+1}\left[s^E(e)\right]\right\}\right] + a_c$$

$$= \log\left[\sum_{e:c(e)=c} \exp\left\{A_{t-1}\left[s^S(e)\right] + \pi_t[b(e); i] + B_{t+1}\left[s^E(e)\right]\right\}\right] + A_c \ (4.44)$$

$$= \max_{e:c(e)=c}^{*} \left\{A_{t-1}\left[s^S(e)\right] + \pi_t[b(e); i] + B_{t+1}\left[s^E(e)\right]\right\} + a_c$$

$$\pi_t(b; e) = \log \left[\sum_{e:u(e)=u} \left\{ \alpha_{t-1}\left[s^S(e)\right] \cdot P_t[c(e); i] \cdot \beta_{t+1}\left[s^E(e)\right] \right\} \right] + a_b$$

$$= \log \left[\sum_{e:u(e)=u} \exp \left\{ A_{t-1}\left[s^S(e)\right] + \pi_t[c(e); i] + B_{t+1}\left[s^E(e)\right] \right\} \right] + A_b \quad (4.45)$$

$$= \max_{e:u(e)=u}^{*} \left\{ A_{t-1}\left[s^S(e)\right] + \pi_t[c(e); i] + B_{t+1}\left[s^E(e)\right] \right\} + a_b$$

where the quantities A_c and A_b are normalization quantities needed to prevent the excessive growth of A and B and the operator max* is defined as

$$\overset{*}{\max_i}(a_1, a_2, \ldots, a_i) = \log \left[\sum_{i=1}^{I} \exp(a_i) \right] \quad (4.46)$$

The major task of the LOG-MAP algorithm is to compute the logarithm of the sum of exponentials, which, in practice, can be approximated as

$$\overset{*}{\max_i}(a_i) = \max_i(a_i) + \delta(a_1, a_2, \ldots, a_I) \quad (4.47)$$

$$\simeq \max_i a_i \quad (4.48)$$

Algorithms using the approximation given in (4.48) are referred to as MAX-LOG-MAP algorithms. Note the following:

- The term $\delta(a_1, a_2, \ldots, a_I)$ in (3.47) is a correction term that can be computed recursively using a 1-D lookup table [152].
- The approximation in (3.48) can be performed without undue performance degradation between medium to high SNRs if $a_{\max} \gg a_i, \forall a_i \neq a_{\max}$, where $a_{\max} = \max_i(a_i)$.
- Using the MAX-LOG-MAP algorithm, good performance can be achieved without undue computational complexity.

4.3   INTERLEAVER DESIGNS FOR TURBO CODES

An innovative feature of turbo codes is the use of a pseudorandom interleaver, which permutes block of bits before encoding. In turbo codes, interleavers are integral part of the coding and decoding strategy itself. Indeed, the high error-correcting power of turbo codes can be traced to three sources:

- the random-like coding introduced by the interleaver-concatenated encoder,
- designing good pseudorandom interleavers with low memory, and
- the closed-loop feedback loop acting around the two-stage decoding receiver.

However, unlike convolutional codes, turbo codes have an error floor at high SNRs. That is, the bit error rate drops very quickly at the beginning, but eventually levels off and becomes flat at high SNRs. This is due the fact that the asymptotic performance characterized by the minimum free distance and the free distances associated with the turbo codes are typically very small. The free distance of a turbo code can be increased by designing interleavers with high "spread." In this context, the interleaver's depth (spread) and storage requirement are challenging issues in the design of turbo codes. Simple structured interleavers can be used to avoid the storage problem as well as to increase the interleaver depth. In particular, for short block lengths on the order of a few hundred bits, structured interleavers offer good performance. The interleavers that have been proposed in the literature include pseudo-random, spread-random, dithered golden interleavers, and dithered relative prime (DRP) interleavers. Among them, the DRP interleavers have the best asymptotic performance, with a very low-memory requirement [29].

4.3.1 Definition of Interleaver Spread

Figure 4.6 shows a definition of interleaver spread. Here, the indexes $1, 2, \ldots, N$ form the input vector of length N to be interleaved, and $\mathbf{A}$ is the vector after interleaving. The spread of the interleaver is defined as

$$S = \min_{i, j, i \neq j} \left[|\mathbf{A}(i) - \mathbf{A}(j)| + |i - j| \right] \tag{4.49}$$

Next, we describe a few interleavers that are commonly used. Pseudorandom interleavers simply permute the elements in a predefined random order. Typically, the pseudorandom interleaver requires N indexes to be stored to implement an interleaver of length of N. Moreover, the interleaver is selected without any restriction and thus may have poor distance properties, causing an error floor problem.

To improve the free distance of a turbo code, spread-random interleavers can be used. In a spread-random interleaver, the permutation order is selected such that any integer in the order is at least S digits away from the previous S integers in the interleaver. The search time to find a spread-random interleaver increases with the spread S, and as a rule of thumb, the choice of $S < \sqrt{N/2}$ could produce a solution within a reasonable time.

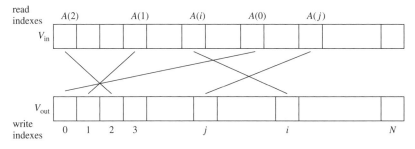

Figure 4.6 Spread of an interleaver.

While random and spread interleavers have a storage problem, structured interleavers can be designed to achieve good minimum distances and low storage requirements. In particular, using a few integer values, unique interleavers can be generated. For example, a simple relative prime (RP) interleaver can be designed using modulo-m arithmetic. For an interleaver length of N, the indices of an RP interleaver are calculated as follows:

$$\mathbf{A}(n) = s + np, \text{modulo} N, n = 0, 1, \ldots, N - 1 \qquad (4.50)$$

where the integers s is the starting index and N and p are relative primes.

Another structured interleaver is the Golden relative prime (GRP) interleaver based on the Golden section. For an interleaver length of N, the indices of the interleaver are calculated using an RP interleaver as given in (4.50) but with the following constraints: in selecting these parameters, the following two conditions must be satisfied to avoid including an index more than once in an interleaver: (1) p and N must be chosen from prime numbers and (2) p must be chosen as the closest prime integer to $c = N(g^m + j)/r$ where g is the Golden section value, $m > 0$ and an integer, r, is the index spacing between nearby elements, and j is any integer modulo r.

Although structured interleavers have good performance for short block sizes, interleavers with some randomness perform better than completely structured interleavers, especially for large block sizes on the order of 1000 bits or more. In this category, the DRP interleavers have the best asymptotic performance and very low-memory requirement [29]. The DRP interleaver design has three stages (see Figure 4.7):

- First, the information bit is locally permuted (also called dithered) using a small read dither vector, $\mathbf{r}$, of length R. The vector $\mathbf{r}$ is a permutation of indices 0 through R, where $R \ll N$. The interleaver will be divided into subvectors consisting of R number of elements, and the subvectors are permuted according to the pattern dither vector $\mathbf{r}$.

- The resulting vector is permuted again using an RP interleaver to obtain good spread.

- Finally, a second local permutation called write dither is permuted using a small dither vector $\mathbf{w}$, of length W, to generate the output vector.

The DRP interleaver can be generated for various block sizes uniquely by storing the dither vectors $\mathbf{r}$ and $\mathbf{w}$ and the relative parameters r and s.

Figure 4.8 plots the performance of an 8-state turbo code with various interleaver designs of size 8192 bits. The figure also shows the performances of an uncoded system and a constraint length 7 convolutional code (64 states) decoded using a Viterbi decoder.[4] As we see from the figure, the DRP interleaver removes the error floor problem associated with the the originally designed Berrou turbo codes (i.e.,

---

[4]We acknowledge Stewart Crozier from CRC for providing this plot.

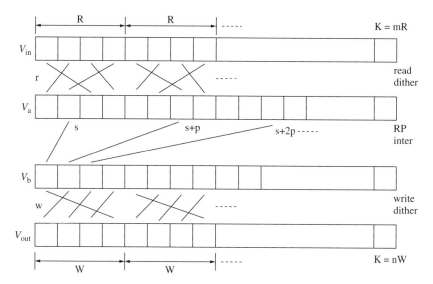

Figure 4.7 Dithered prime interleaver (DRP).

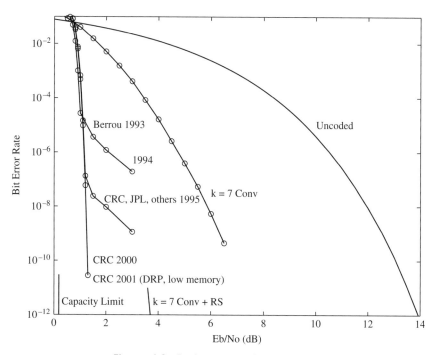

Figure 4.8 Performance of turbo codes.

random interleavers [17]) and RP and GRP interleavers. Moreover, at a bit error rate level of 10^{-9}, gains of 6 dB and 12 dB are achieved over the convolutionally coded and uncoded systems, respectively. Figure 4.8 also shows the capacity limit of the particular scenario and the performance with concatenated convolutional and Reed-Solomon codes.

Furthermore, a significant reduction in storage space can be achieved by using DRP interleavers, particularly when a bank of interleavers is needed, as in the following communication example.

4.4   SPACE-TIME TURBO CODES

Turbo codes have closed the gap between the theoretical capacity limits and the practically achievable information rates for Gaussian SISO channels. The new challenge is to close this gap for more powerful communications systems such as multi-element antenna systems. Several versions of modified turbo codes for MIMO schemes with two transmit antennas, based on parallel concatenated turbo codes, have been proposed in the literature, namely, space-time turbo codes. The space-time turbo codes proposed in [157] and [145] can be classified into three classes, as shown in Figures 4.8–4.11:

- In the first class (Figure 4.9), the outputs of a turbo code are transmitted using multiple antennas.
- In the second class (Figure 4.10), the outputs of a turbo code are bit interleaved, mapped to QPSK/QAM symbols, and transmitted using multiple transmit antennas.
- In the third class (Figure 4.11), the transmitter is composed of a turbo encoder followed by the operations of puncturing,[5] channel interleaving, and multiplexing.

Multiple transmit antennas are used to transmit the output of the multiplexer. However, the generalization of these algorithms for more than two transmit antennas is not obvious.

4.4.1   Example Space-Time Turbo Codes

The encoder of this space-time turbo code is an 8-state turbo code followed by the operations of puncturing, multiplexing, channel interleaving, and modulation. The multiplexing and channel interleaving take advantage of space and time diversity, respectively. Different information rates are achieved by using different puncturing patterns. In Figure 4.12, we show the puncturing pattern used for a rate 1/2 turbo code. In this scheme, one of the transmit antennas sends the punctured parity 1 and

---

[5]Puncturing refers to deleting certain parity check bits, thereby increasing the data rate.

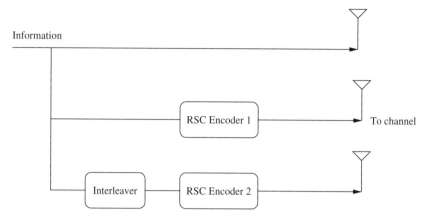

Figure 4.9 Space-time turbo encoder 1.

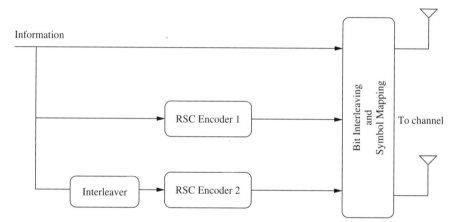

Figure 4.10 Space-time turbo encoder 2.

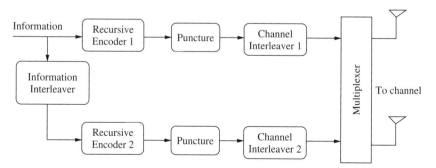

Figure 4.11 Space-time turbo encoder 1.

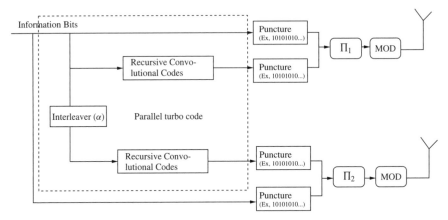

Figure 4.12 Transmitter block diagram.

part of the information data, and the other transmit antenna sends the rest of the information bits and the punctured set of parity 2, as symbols drawn from a QAM constellation. We use rate-compatible punctured turbo codes to achieve various information rates.

With no delay spread, the discrete-time model of the received signal at the ith time instant is given by

$$\mathbf{y}(i) = \mathbf{H}(i)\mathbf{x}(i) + \mathbf{v}(i) \tag{4.51}$$

where $\mathbf{H}$ is the 2×2 complex channel matrix, $\mathbf{x}$ is the 2×1 information-bearing vector, $\mathbf{y}$ is the 2×1 received signal vector, and $\mathbf{v}$ is a 2×1 Gaussian noise vector. The components of the noise vector are uncorrelated complex white Gaussian random variables with zero mean and variance $N_0/2$ for each dimension. The elements of the channel matrix are assumed to be flat Rayleigh fading with unit variance and zero mean.

Figure 4.13 illustrates the receiver structure, where we separate the optimal decoding problem into two stages, inner and outer decoding, and exchange all information learned from one stage to another iteratively until the receiver converges. The inner and outer stages are separated by the interleavers Π_1 and Π_2 and the de-interleavers Π_1^{-1} and Π_2^{-1} to decorrelate the correlated outputs before feeding them to the next decoding stage. The de-interleaver is used to compensate for the interleaving operation used in the transmitter. The iterative receiver produces new and hopefully better estimates at each iteration and repeats the information exchange process a number of times to improve the decisions.

The outer decoder of the iterative algorithm is a turbo decoder based on the SISO decoder implemented using the generalized BCJR algorithm explained in this chapter. The decoder has two iterative loops that operate simultaneously, as shown by the arrows in Figure 4.13. We use a suboptimal parallel interference cancellation (PIC) detector for the inner decoder, which is based on the minimum

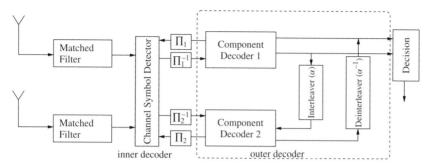

Figure 4.13 Receiver block diagram. The interleavers α and Π are associated with the turbo codes and the channel, respectively.

mean-square error (MMSE) principle and soft interference cancellation. PIC optimizes the interference estimate and the weights of the linear detector jointly, as explained in Chapter 5. We estimate the expectations of interfering substreams using the *a priori* probabilities of the transmitted bit streams provided by the SISO channel decoders at the previous iteration.

Simulations Simulation results are presented for QPSK transmission over an ergodic two-input, two-output channels using 8-state turbo codes which are punctured to achieve various rates. The performance measures are the bit error rate (BER) and packet error rate (PER) under the assumption of an ideal Rayleigh channel communication scenario. For all the simulations, we consider dual-terminated turbo codes that are composed of two identical 8-state convolutional codes generated by using the recursive FB generator polynomial [1011] and the FF polynomial [1101]. Various code rates are obtained by rate-compatible puncturing. We consider random and designed dithered random (DRP) interleavers (shown by α in Figures 4.12 and 4.13). The transmission is organized in blocks of information bits of 3200. The selected DRP interleavers have a high minimum measured distance, so the error rate floor is both low and steep. Moreover, the DRP interleavers can be stored and implemented using a small number of parameters, which makes them suitable for designing large interleaver banks [29]. Decoding is stopped early (using fewer than 30 iterations) if a codeword pattern recurs for three consecutive iterations. The total power P is divided equally into each antenna.

Figures 4.14 and 4.15 show, respectively, the BER and PER performances of the STTC for a system with $M = N = 2$. In the figures, we show the performance of the turbo codes designed with random and DRP interleavers. In all of the simulations, the information rate per transmission is calculated as

$$\text{bits per modulation} \times \text{average code rate} \times \text{number of antennas}$$

In the case of two transmit antennas and QPSK modulation with rates $R = 1/3$, 1/2, 2/3, and 3/4, the information rate per transmission = 1.33, 2, 2.66, and 3 bits per channel use, respectively.

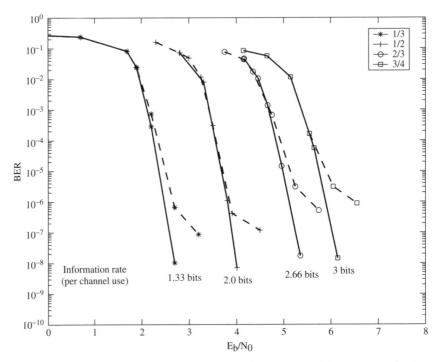

Figure 4.14 BER performance of space-time turbo codes with two transmit and two receive antennas. Solid curves: STTC with DRP interleavers; dashed curves: STTC with random interleavers.

From the figures, we see that with DRP interleaving, the space-time turbo code performs well at low BER and PER levels and the four systems operate 4, 2.8, 2.2, and 2 dB away from their respective capacity limits at 10^{-8} BER and 10^{-5} PER. Although the proposed scheme does not come close to the capacity limits of multiple-antenna systems, the throughput gain is still much larger than that of a SISO system operating in an AWGN channel. Moreover, as we can see from Figures 4.14 and 4.15, the performance improvement of DRP over random interleaving is significant at low BER and PER levels.

4.5 MULTIRATE LAYERED SPACE-TIME (MLST) TURBO CODES

Next, simulation results are presented for QPSK transmission over the quasi-static MIMO channels using MLST turbo codes with various rates. We refer the readers to Section 3.6 for the details of MLST turbo codes.

The performance measures are the BER and PER. The $n_t \times n_r$ matrix channel **H** remains constant within a packet and changes randomly for every packet. The total power P is divided equally among the transmit antennas. That is, P/n_t power per layer in a symbol interval.

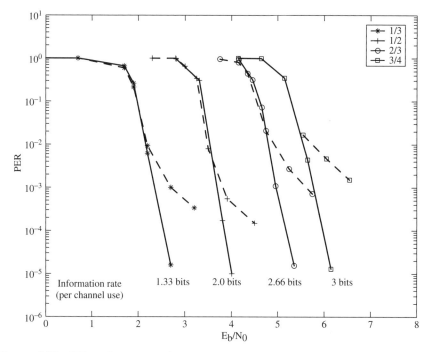

Figure 4.15 PER performance of space-time turbo codes with two transmit and two receive antennas. Solid curves: STTC with DRP interleavers; dashed curves: STTC with random interleavers.

For all the simulations, the turbo codes are composed of two identical 8-state convolutional codes generated by using a recursive FB generator polynomial [1011] and an FF polynomial [1101]. Various code rates are obtained by puncturing the coded bits of the rate 1/3 turbo codes. No effort is made to optimize the pseudo-random interleavers. In all the examples, the transmission is organized in blocks of 8012 coded bits; that is, the block size of a turbo code is fixed to 8012 coded bits per antenna regardless of its rate. Thus, to achieve various coding rates, we change the number of information bits (and the interleaver size) transmitted by an antenna. The maximum number of iterations is restricted to 20. Decoding is stopped early (using fewer than 20 full iterations) if a codeword pattern recurs for three consecutive iterations [59]. In the interference cancellation step, the errors in the previously decoded symbols are propagated in the subsequent decodings. The per-layer rates are chosen based on the empirically estimated 1% outage layer rates.

In the following, we show the PER versus SNR performances for the following schemes operating in quasi-static fading channels: (1) a multirate system with DSTI, (2) a multirate system without DSTI, and (3) V-BLAST. Figures 4.16(a) and 4.16(b) illustrate the per-layer PER and average PER versus SNR performances for $M = N = 4$, respectively. In Figure 4.16(a), we show the PERs of each layer separately.

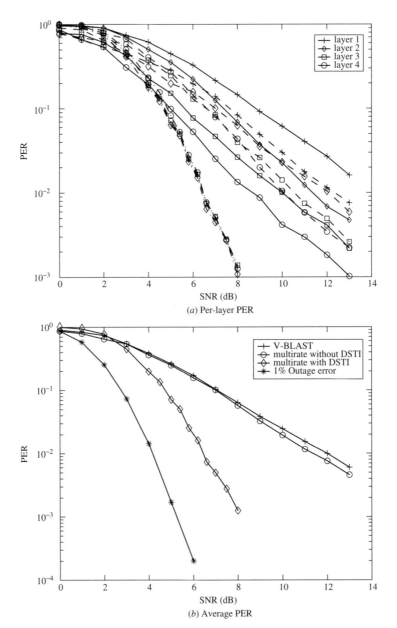

Figure 4.16 (*a*) Per-layer PER and (*b*) average PER versus SNR performances for a 4 × 4 system with 8-state turbo codes. (*a*) Solid curves represent the performance of V-BLAST-OSIC with rate 0.5 codes used in all four antennas; dashed curves represent the performance of a multirate (without DSTI) coded system with layer code rates (0.3, 0.4, 0.5, 0.8); and dashed-dot curves represent the performance of multirate system with DSTI with layer code rates (0.42, 0.46, 0.52, 0.6). (*b*) The corresponding PER averaged over the performances of all four antennas.

The average PERs taken over all four antennas are shown in Figure 4.16(b). We also show the 1% outage capacity results. As expected, scheme (1) performs better than the other two schemes. Different layers for multirate systems with DSTI have roughly the same error performance, while for V-BLAST or multirate systems without DSTI, different layers experience different error rates. In the multirate system with DSTI, each layer is transmitted through all possible channels and is not affected by the fading of any particular channel. Also, by design, each layer has the same outage probability. However, in the multirate scheme without DSTI, the per-layer PER is affected by deep fading of the channels. In V-BLAST, we consider the case where every antenna transmits its own independently encoded horizontal layer of data with equal-rate codes and the receiver performs OSIC detection and decoding, descrbed in Subsection 3.3.1. One possible shortcoming of V-BLAST is that its overall performance may be dominated by the weakest layer, particularly the first decoded layer, because it has the lowest diversity in V-BLAST decoding. Without error propagation, the first detection layer has the minimum SINR among the n_t possible layers, since it faces $n_t - 1$ interferences; the second detection layer has only $n_t - 2$ interferences; and so on. Accordingly, the per-layer PERs are significantly different from each other and the first decoded layer has the worst PER, followed by the second decoded layers, the, third decoded layers, and so on. In the OSIC detection scheme, the optimal order of detection has to be determined at each decoding step, which is computationally expensive. For the proposed multirate architectures, we use a predetermined detection order assigned on the basis of the layer coding rates. Thus, the complexity is lower than that of the the the V-BLAST OSIC receiver.

Moreover, in Figure 4.17, we show the BER and PER performance in the case where $n_t = n_r = 6$. As we see, the larger the n_t and n_r, the higher the performance gain using scheme (1) compared to the other two schemes. Moreover, at 1% outage, the performance of the proposed multirate DSTI codes is only about $1-2$ dB away from the capacity limits and gains about 5 dB over V-BLAST.

4.6 SUMMARY AND DISCUSSION

In this chapter, we discussed the turbo coding principle, focusing on the techniques that are of special interest to MIMO wireless communications. In particular, the application of the turbo coding principle in the MIMO framework, the so-called space-time turbo codes for improved diversity and coding gain performance, were discussed. Much of the discussion, however, was devoted to the following issues:

- serial and parallel concatenated turbo codes and their iterative decoders,
- SISO modules, which are exemplified by the BCJR algorithm that performs maximum *a posteriori* estimation on a symbol-by-symbol basis, and
- the extraction of extrinsic information which we believe is behind the success of turbo principles.

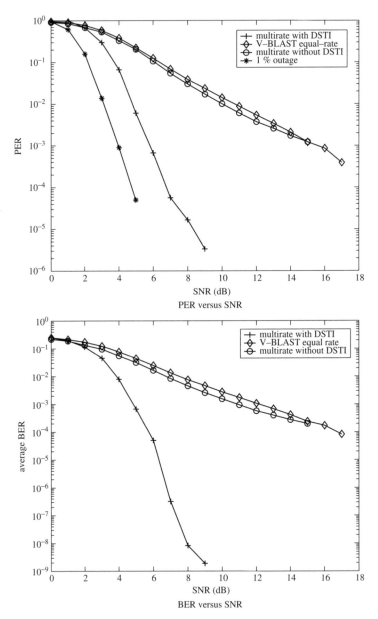

Figure 4.17 PER/BER versus SNR performances for a 6 × 6 system with 8-state turbo codes. The solid-diamond curve represents the performance of V-BLAST-OSIC with rate 0.5 codes used in all six antennas. The solid-circle curve represents the performance of a multirate (without DSTI) coded system with layer code rates (0.35, 0.39, 0.48, 0.52, 0.57, 0.72). The solid-plus curve represents the performance of a multirate system with DSTI with layer code rates (0.38, 0.44, 0.48, 0.52, 0.57, 0.6).

In 1948, Shannon laid down the foundation of information theory. In particular, he established the upper limit of the data transmission rate over a given channel and described channel coding as the means for achieving this limit: *For any communication channel there exist families of random block codes that achieve an arbitrarily small probability of error at any information rate up to the capacity of the channel* [137]. Although Shannon's proof advocated the use of random codes, it did not provide any practical coding and decoding algorithms to achieve the limits. Since then, we have been challenged by a fundamental research question: *How can we practically approach the Shannon channel capacity using random block codes?* In practice, the key obstacle to approaching the channel capacity is not the construction of good long random codes; rather, it is how to keep the decoding complexity reasonable. Serial concatenation of the Reed-Solomon outer code [121] followed by a convolutional inner code [160] was the most popular concatenated coding scheme until the new family of convolutional codes known as turbo codes were invented by a group of researchers in France in 1993 [16, 17]. Turbo codes are built from either parallel or serial concatenation of recursive systematic convolutional codes linked together by nonuniform interleaving.

Typically, the concatenation of codes splits the decoding problem into manageable steps. The crucial innovation of turbo decoding consists not only of splitting the burden of decoding into steps but also of passing what has been learned from previous steps and doing so iteratively. The discovery of turbo codes and iterative decoders also rekindled interest in low-density parity-check (LDPC) codes [51].

Turbo encoding-decoding has become popular due to its simple structure, which permits relatively low-complexity iterative decoding, yet performs within 0.7 dB of the theoretical Shannon capacity limits at BERs of approximately 10^{-5}. It is amazing to find that both turbo codes ([16], [17]) are the closest known codes to the capacity limits with reasonable decoding complexity and have a random code-like weight distribution. Most importantly, iterative decoders allow both of these codes to achieve their error performance within a hair's breadth of Shannon's theoretical limit on the channel capacity in a physically realizable fashion.

5

TURBO-BLAST

5.1 INTRODUCTION

Turbo-BLAST (T-BLAST) is a novel MIMO antenna scheme for high-throughput wireless communications. It exploits the following three ideas:

1. The BLAST architecture.
2. The random layered space-time (RLST) coding scheme by using independent block codes and random space-time interleaving.
3. The suboptimal turbo-like receiver for decoding the RLST codes and estimating the channel matrix in an iterative and, most important, simple fashion.

The net result is a new transceiver that is not only computationally efficient compared to the optimal maximum likelihood decoder, but also yields a probability of error performance that is orders of magnitude smaller than that of V-BLAST for the same operating conditions. This chapter also presents experimental results using real-life indoor channel measurements, demonstrating the high-spectral efficiency of T-BLAST.

5.2 T-BLAST: BASIC TRANSMITTER CONSIDERATIONS

Consider a MIMO system that has n_t transmitting and n_r receiving antennas. Throughout the chapter, we assume that (1) the n_t transmitters operate with

Space-Time Layered Information Processing for Wireless Communications,
By Mathini Sellathurai and Simon Haykin
Copyright © 2009 John Wiley & Sons, Inc.

synchronized symbol timing at a rate of $1/T_s$ symbols per second and (2) the sampling times of n_r receivers are symbol-synchronous. The channel variation is assumed to be negligible over M symbol periods comprising a packet of symbols. Moreover, we only consider a narrowband frequency-flat communication environment, i.e, no delay spread. The extension of this scheme to a frequency-selective environment is straightforward.

Figure 5.1 shows a high-level description of the T-BLAST architecture. The encoding process involves

- Demultiplexing the user information bits into n_t substreams $\{\mathbf{b}_k\}_{k=1}^{n_t}$ of equal data rate.
- Independent block encoding of each data substream, which uses the same predetermined linear block FEC code with a minimum weight equal to $d_{\min}$, as shown by

$$\mathbf{C} = [\mathbf{b}_1\mathbf{G}, \mathbf{b}_2\mathbf{G}, \ldots, \mathbf{b}_{n_t}\mathbf{G}] = [\mathbf{c}_1, \mathbf{c}_2, \ldots, \mathbf{c}_{n_t}]^T \qquad (5.1)$$

where $\mathbf{G}$ is the $K \times L$ binary code generator, the $\{\mathbf{b}_k\}_{k=1}^{n_t}$ are K-dimensional information sequences, and the $\{\mathbf{c}_k\}_{k=1}^{n_t}$ are L-dimensional code sequences.
- Bit-interleaved encoded substreams using a random space-time permuter Π. We use $\tilde{\mathbf{C}} = \{\tilde{\mathbf{c}}_k\}_{k=1}^{n_t}$ to denote the permuted substreams:

$$\tilde{\mathbf{C}} = \Pi(\mathbf{C}) \qquad (5.2)$$

The random space-time interleaver is independent of the incoming data streams, and its design must guarantee the use of an entire subchannel by each independently coded substream in an equal manner, thereby permitting the use of an off-line design procedure. In the rest of the chapter, we consider a space interleaver based on diagonal layering of each independently coded substream, which is followed by random time interleavers to generate the RLST codes, as shown in Figure 5.2. The space interleaving procedure is

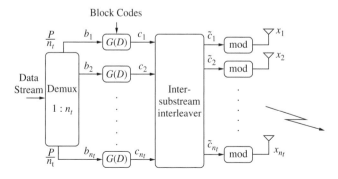

Figure 5.1 T-BLAST transmitter.

c_1	c_6	c_5	c_4	c_3	c_2	c_1	c_6	c_5	c_4	c_3
c_2	c_1	c_6	c_5	c_4	c_3	c_2	c_1	c_6	c_5	c_4
c_3	c_2	c_1	c_6	c_5	c_4	c_3	c_2	c_1	c_6	c_5
c_4	c_3	c_2	c_1	c_6	c_5	c_4	c_3	c_2	c_1	c_6
c_5	c_4	c_3	c_2	c_1	c_6	c_5	c_4	c_3	c_2	c_1
c_6	c_5	c_4	c_3	c_2	c_1	c_6	c_5	c_4	c_3	c_2

Figure 5.2 Diagonal space interleaver.

simply a permutation over the L columns according to the interleaver. Note that unlike D-BLAST, T-BLAST does not experience any boundary wastage in its diagonal layering structure due to the *cyclic* nature of the encoding process.

- The space-time interleaved substreams are independently mapped onto symbols $\mathbf{X} = \{\mathbf{x}_k\}_{k=1}^{n_t}$, where

$$\mathbf{X} = f(\tilde{\mathbf{C}}) \tag{5.3}$$

Each interleaved substream is transmitted using a separate antenna. The transmitted signals are received by n_r receiving antennas, whose output signals are fed to an iterative decoding receiver.

5.2.1 Space-Time Interleaving

We propose two types of random space-time interleaving schemes for the T-BLAST architecture:

1. A random intersubstream interleaving of size $n_t \times L$.
2. A space-time interleaving made up of two stages:
   - Stage 1: time interleaving using n_t different and independent random permuters of size L.
   - Stage 2: space interleaving using L different and independent random permuters of size n_t.

Diagonal Layered Space Interleavers The space interleaving proposed in T-BLAST is used for spatial cycling of each substream over all possible subchannels. A deterministic space interleaver based on diagonal layering of each independently coded substream is shown in Figure 5.3. The interleaving procedure is simply a permutation over the L columns in accordance with the design of

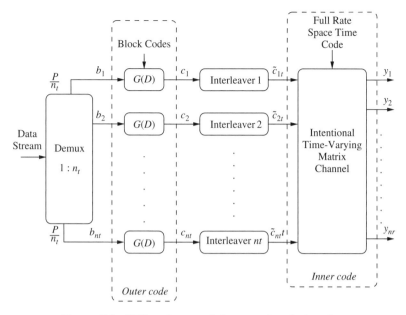

Figure 5.3 RLST codes as serially concatenated codes.

the interleaver. The benefits of T-BLAST using diagonal space-time interleavers include the following:

- Diagonal layering space interleaving guarantees the equal use of each sub-channel by each independently coded substream.
- Each substream can be coded using 1-D code blocks of equal rates.

Without explicit knowledge of the channel matrix at the transmitter end, this may be the best we can achieve for any set of subchannels.

5.2.2 Intentional Time-Varying Channel

The combined use of block codes and interleaving provides the basis for the *random block codes*, namely, parallel and serially concatenated turbo codes. Using this principle for MIMO systems, we construct RLST block codes by concatenating block encoders and space-time interleavers.

Figure 5.3 illustrates another view of the proposed turbo space-time block codes under a quasi-static Rayleigh fading environment. In this representation, we include the effect of diagonal space-time interleaving with the quasi-static Rayleigh matrix channel. Then diagonal interleaving followed by modification of the quasi-static Rayleigh matrix channel introduces an intentional time-varying channel.

The channel shown in Figure 5.4(a) is generated for a (16,16)-BLAST system; note that each subchannel is static within each packet. In Figure 5.4(b), we show the time-varying subchannels generated by the intersubstream interleaving process, that is, by combining the space-time interleaver and the channel shown in Figure 5.4(a). Only 3 subchannels (out of 16) are shown here for simplicity. For a sufficiently large number of transmitters, a fast time-varying channel can be achieved even in delay-limited and nonergodic systems. Moreover, the time averages of each independent channel in Figure 5.4(b) will approach their corresponding ensemble averages in the limit as the observation interval T and the number of transmit antennas n_t approach infinity; thus, the space-time interleaver generates an artificially generated ergodic process from the nonergodic quasi-static Rayleigh-fading MIMO channels. Note also that in Figure 5.4(a), each subchannel is nonergodic since it does not change with time.

5.3   OPTIMAL DETECTION

With no delay spread, the discrete-time model of the received signal is defined by

$$\mathbf{Y} = \sqrt{\frac{\rho}{n_t}}\tilde{\mathbf{H}}\mathbf{X} + \mathbf{V} \tag{5.4}$$

where $\tilde{\mathbf{H}} \in \mathbb{C}^{n_r \times n_t}$ is the known intentional time-varying channel matrix. $\mathbf{X} \in \mathbb{C}^{n_t \times N_s}$ is the transmitted information matrix, $\mathbf{Y} \in \mathbb{C}^{n_r \times N_s}$ is the corresponding received signal matrix, and $\mathbf{V} \in \mathbb{C}^{n_r \times N_s}$ is a noise matrix. The SNR received at each receiving antenna is denoted by ρ. The components of the noise matrix are uncorrelated zero-mean complex white Gaussian random variables with zero mean and variance σ^2.

The optimal receiver performs an exhaustive search to determine $\mathbf{X}$ from the received signal $\mathbf{Y}$ in accordance with the formula

$$\hat{\mathbf{X}} = \arg\min_{\mathbf{X}} \left\| \mathbf{Y} - \sqrt{\frac{\rho}{n_t}}\tilde{\mathbf{H}}\mathbf{X} \right\|^2 \tag{5.5}$$

The computational complexity of this search increases exponentially with the number of transmit antennas n_t and the number of information bits in the modulation and the block size N_s.

In theory, it is possible to model the proposed RLST code as a single Markov process, and a trellis can be formed to include the effect of interleaving. Such a trellis representation is extremely complex and does not lend itself to feasible decoding algorithms [91].

5.4   DISTANCE SPECTRUM OF RLST CODES

We assume the use of the same 1-D codes with minimum distance $d_{\min}$ to encode each substream separately. In the case of convolutional codes, $d_{\min}$ will be the free distance d_{free} of the code. We consider binary phase-shift keying (BPSK) modulation for data transmission.

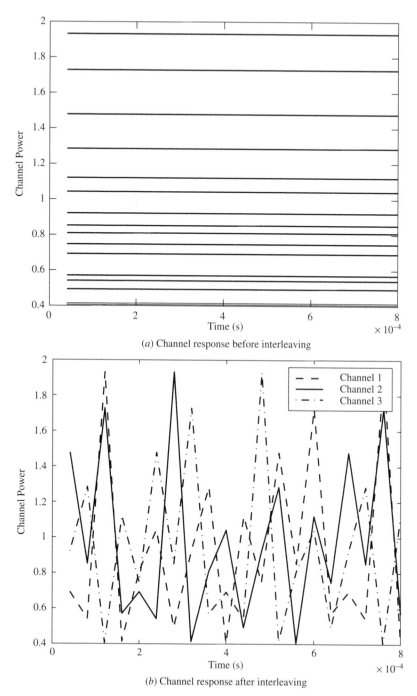

(*a*) Channel response before interleaving

(*b*) Channel response after interleaving

Figure 5.4 Intentional time-varying channel.

Assuming quasi-static fading, we define the noiseless matrix representation of the received signal as

$$S(X) = \sqrt{\frac{1}{n_r}} HX \tag{5.6}$$

We further assume that the entries $H_{i,j}$ of the $n_t \times n_r$ channel matrix H satisfy the condition $\mathcal{E}\|H_{i,j}\|^2 = 1$, where $\mathcal{E}$ is the expectation operator. The normalization factor $1/\sqrt{n_r}$ is introduced in (5.6) to ensure that the total power received from each transmit antenna remains constant as n_r varies. $X \in \mathbb{C}^{n_t \times N_s}$ is the transmitted encoded signal matrix. The asymptotic probability of error, in the limit as the receiver AWGN goes to zero, is determined by the minimum Euclidean distance between any pair of code sequences X and $\tilde{X}$. The squared minimum distance of the RLST code is defined by

$$d^2_{ST,min}(X, \tilde{X}) = \min_{X,\tilde{X}} \left\| \frac{S(X) - S(\tilde{X})}{2} \right\|^2 \tag{5.7}$$

This squared distance is equivalent to

$$d^2_{ST,min}(X, \tilde{X}) = \min_{e \neq 0} \sum_{l=1}^{N_s} e(l)^T \Delta e(l) \tag{5.8}$$

where $e(l) \in \{-1, 0, +1\}^{n_t}$ is the error vector defined as $\frac{x_l - \tilde{x}_l}{2}$, where x_l is lth column of matrix X, and the random matrix $\Delta = (1/n_r)H^\dagger H$ is defined by

$$\Delta = \begin{bmatrix} 1 & \delta_{1,2} & \cdots & \delta_{1,n} \\ \delta_{2,1} & 1 & \cdots & \delta_{2,n} \\ \vdots & \vdots & \ddots & \vdots \\ \delta_{n,1} & \delta_{n,2} & \cdots & 1 \end{bmatrix}, \quad \delta_{i,j} < 1, \forall i, j \tag{5.9}$$

In the sequel, we reveal the importance of the random interleavers in the RLST code design using 1-D channel codes. The decision distance d^2_{ST} of RLST codes simplifies to

$$d^2_{ST} = \sum_{l=1}^{N_s} e(l)^T \Delta e(l)$$

$$= \sum_{l=1}^{N_s} \left\{ \sum_{j=1}^{n_t} e_j^2(l) + 2 \sum_{i<j} \delta_{ij} e_i(l) e_j(l) \right\} \tag{5.10}$$

where $e_j(l)$ is the jth element of the vector $e(l)$. For the special case of $n_t = 1$, the minimum distance $d_{ST,min}$ simplifies to d_{min}. However, for $n_t > 1$, it is not trivial to preserve the minimum distance when we use 1-D linear block codes as substream codes.

By definition, we have $\delta_{ij} < 1$ for $i \neq j, \forall i, j$. Consequently, the lower bound on the minimum distance of RLST code becomes

$$d_{ST,\min}^2 > \sum_{l=1}^{N_s} \left\{ \sum_{j=1}^{n_t} e_j^2(l) - 2 \sum_{i<j} |(e_i(l)| \times |e_j(l)|) \right\} \qquad (5.11)$$

When only v substreams have nonzero error events of weight $d_{\min}$ among the n_t possible substreams, (5.11) reduces to

$$d_{ST,\min}^2 > \sum_{l=1}^{N_s} \left\{ \sum_{j=1}^{v} e_j^2(l) - 2 \sum_{i<j} |e_i(l)| \times |e_j(l)| \right\} \qquad (5.12)$$

Proposition 1 Without interleaving: *The minimum distance of an RLST code is dependent on the 1-D channel codes used by each substream. When an RLST code is designed using existing 1-D linear block codes, the minimum distance of such a code will be lower bounded by zero. In particular, the minimum distance will be zero when there is an even number of nonzero error events occurring in the RLST code and pairs of error events are aligned in time such that they are the negative of each other. This is because the sum of any two codewords is a permissible codeword, which is a fundamental property of linear block codes.*

Proposition 2 With interleaving: *Consider the time-interleaving operation before the space interleaving so that we may include the space interleaving with the channel matrix to get an intentional time-varying channel. In this case, we can align an error event with another, only at one bit interval, by independent random time interleaving of each substream separately. The probability of aligning more bits is a function that depends on the length of the interleavers. For random interleaving, we have the following probabilitites ([91], [92]):*

$$P\{ |\mathbf{e}_j(l)| = 1 \} = \frac{d_{\min}}{N_s}$$

$$\qquad (5.13)$$

$$P\{ |\mathbf{e}_j(l)| = 0 \} = 1 - \frac{d_{\min}}{N_s}$$

When interleaver length $N_s \to \infty$, with any choice of a random interleaver we have $d_{ST,\min} \geq d_{\min}$. For finite N_s, we show that the expected value of the minimum distance of RLST codes, where the expectation is taken over all random interleavers of length N_s, can be made equal to the minimum distance of the 1-D linear channel codes for sufficiently large N_s. This is a sufficient condition to show the existence of random interleavers needed to design an RLST code that preserves the minimum distance of 1-D linear codes.

The expectation, over all random interleavers of length N_s, of the minimum distance of RLST code is given by

$$E\left\{d_{ST,\min}^2\right\} > \sum_{l=1}^{N_s}\left\{\sum_{j=1}^{v} \mathbf{e}_j^2(l) - n_t(n_t-1)\frac{d_{\min}^2}{N_s^2}\right\} - 2v$$

$$= vd_{\min} - n_t(n_t-1)\frac{d_{\min}^2}{N_s} - 2v$$

(5.14)

where $v > 0$ represents the ability to align v bits of v substream error events. To achieve the minimum distance $d_{\min}$ of 1-D linear block codes, we should have $E\{d_{ST,\min}^2\} > d_{\min}$. Given that the length of the interleaving must be

$$N_s > \frac{n_t(n_t-1)d_{\min}^2}{(v-1)d_{\min} - 2v}$$

(5.15)

the minimum of this condition is achieved for $v = 2$ and $d_{\min} > 4$. Therefore, a sufficient condition to preserve the minimum distance property of the RLST code is the interleaver length N_s that must satisfy the condition

$$N_s > n_t(n_t-1)d_{\min}^2$$

(5.16)

where $d_{\min} > 4$.

This means that, with random-time interleaving, even in the presence of channel correlations, the minimum distance property of 1-D channel codes can be preserved in MIMO antenna schemes, provided that the interleavers are carefully chosen and their sizes are sufficiently large. The possible presence of bad interleavers is rare when we use large random interleavers. Consequently, by using RLST codes with large interleavers and ML decoding, the asymptotic error performance of a MIMO system may be made equivalent to that of a single transmit antenna system in AWGN channels using the same total power P. A similar analysis can be found in [94] and [95] for multiuser detection using iterative decoders in AWGN channel. Moreover, we note the following:

1. If the entries $H_{i,j}$ of $n_t \times n_r$ channel matrix $\mathbf{H}$ are independent, then $E\|H_{i,j}\|^2 = 1$. For fixed n_t, by the law of large numbers, we find that the random matrix $\Delta \to \mathbf{I}_{n_t}$ almost surely as n_r gets large, where $\mathbf{I}_{n_t}$ is the $n_t \times n_t$ identity matrix. Thus, each substream is decoupled independently with a guaranteed minimum distance of $d_{ST,\min} = d_{\min}$.

2. The result described above is valid not only for a quasi-static channel but also for the time-varying fading case. In the latter channel, the random matrix Δ is time-varying. For large n_r, no additional gain may be obtained. However, for finite n_r, time diversity due to the time-selective fading will provide additional coding gain for both correlated and uncorrelated channel environments.

3. As in the time-varying channel case, diagonal space interleaving provides no additional gain for large n_r. However, in the case of finite n_r, space interleaving plays a major role in providing additional freedom for interleaving depth

as well as intentional time-selective fading for each decoder to achieve the optimal performance on each substream. For finite n_r, the diagonal elements of the random matrix Δ, $\delta_{ii} \leq 1$ for all i, and space interleaving is necessary to guarantee equal use of the entire channel by each substream.

5.5 ITERATIVE DECODING: BASIC CONSIDERATIONS

As we have seen, the optimal signal-decoding problem in intersubstream encoded MIMO systems has a computational complexity that is exponential in the number of substreams, the constellation size, and the block size. Even though it is possible to model the proposed RLST block code as a single Markov process and a trellis can be formed to include the effect of space-time interleaving, optimal decoding of such a trellis representation is extremely complex and does not lend itself to feasible decoding algorithms [94].

This section proposes a practical suboptimal detection scheme based on iterative turbo detection principles. The intersubstream coding proposed as independent encoding and space-time interleaving can be viewed as a serially concatenated code, as illustrated in Figure 5.5:

- Outer code: n_t parallel channel codes
- Inner code: time-varying channel matrix

The inner and outer codes are separated by n_t parallel interleavers. The concatenated code can be decoded using a lower-complexity iterative receiver similar to the iterative schemes proposed for serially concatenated turbo codes. In the iterative decoding scheme, we separate the optimal decoding problem into two stages and exchange the information learned from one stage to the other stage iteratively until the receivers converge. The two decoding stages are:

- Inner decoding: SISO channel estimation
- Outer decoding: a set of n_t parallel SISO channel decoding

The detector and decoder stages are separated by space-time interleavers and de-interleavers. The interleavers and de-interleavers are used to compensate for the interleaving operation used in the transmitter as well as to decorrelate the correlated outputs before feeding them to the next decoding stage. The iterative receiver produces new and better estimates at each iteration of the decoding process and repeats the information-exchange process a number of times to improve the decoding decisions and channel estimates. Note that the design of our intersubstream coding uses independent coding of each substream; hence, the receiver needs to select only one of 2^{N_s} sequences for each n_t sequence separately without increasing the probability of symbol error significantly. The iterative decoder is shown in Figure 5.5.

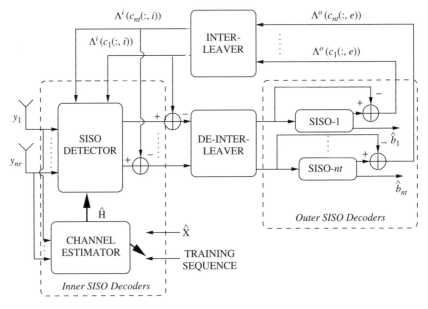

Figure 5.5 Iterative decoder.

5.5.1 Iterative Decoding Algorithm

The iterative decoding structure of serially concatenated turbo codes provides the principal model for the iterative decoding algorithm. The following notations are used to explain the algorithm: the log-likelihood ratios (LLRs) Λ^i and Λ^o, with superscripts i and o, denote the LLR associated with the inner decoder and outer decoder, respectively. The symbols $\Lambda(:, i)$, $\Lambda(:, e)$, and $\Lambda(:, p)$ at the output and input of the SISO modules refer to intrinsic, extrinsic, and *a posteriori* information formulated as LLRs.

First, we define the *a posteriori* LLR of a transmitted bit symbol $c_j(l)$, $j = 1, 2, \ldots n_t$ and $l = 1, 2, \ldots, N_s$:

$$\Lambda(c_j(l); p) = \log \frac{P\left\{c_j(l) = +1 | \mathbf{r}\right\}}{P\left\{c_j(l) = -1 | \mathbf{r}\right\}} \tag{5.17}$$

Using Bayes' rule, (5.17) can be rewritten in the equivalent form

$$\Lambda\left(c_j(l); p\right) = \log \frac{P\left\{\mathbf{r}(t) | c_j(l) = +1\right\}}{P\left\{\mathbf{r}(t) | c_j(l) = -1\right\}} + \log \frac{P\left\{c_j(l) = +1\right\}}{P\left\{c_j(l) = -1\right\}}$$

$$= \Lambda(c_j(l); e) + \Lambda(c_j(l); i) \tag{5.18}$$

The first term $\Lambda(c_j(l); e)$ in (5.18) constitutes extrinsic information and the second term $\Lambda(c_j(l); i)$ constitutes intrinsic information, both pertaining to the code bit $c_j(l)$.

Figure 5.5 illustrates the iterative decoder and depicts message passing between the inner/detector and outer/decoder SISO modules:

1. The SISO detector (inner SISO module) generates soft estimates of the code bits $c_j(l)$ conditioned on the received signal $\mathbf{r}(t)$, as well as the intrinsic information about all the code bits $c_k(l)$, $\forall k, k \neq j$, and $c_j(t)$, $\forall t, t \neq l$. The soft information on $c_j(l)$, computed by the SISO detector, is influenced by the intrinsic information of $\Lambda(c_j(l); i)$ from the previous stage. Specifically, the computation proceeds as follows:
   - Estimate the *a posteriori information*

$$\Lambda^i(c_j(l); p) = \log \frac{P\left\{c_j(l) = +1 | \mathbf{r}, \Lambda^i(\mathbf{C}; i)\right\}}{P\left\{c_j(l) = -1 | \mathbf{r}, \Lambda^i(\mathbf{C}; i)\right\}}, \forall j, l \qquad (5.19)$$

During the first iteration, the initial intrinsic probabilities of all symbol bits are assumed to be 1/2 (i.e., equally likely). Thus, $\Lambda(\mathbf{c}_j(l); i) = 0, \forall j, l$.
   - Compute the extrinsic information

$$\Lambda^i(\mathbf{C}; e) = \Lambda^i(\mathbf{C}; p) - \Lambda^i(\mathbf{C}; i) \qquad (5.20)$$

where $\Lambda^i(\mathbf{C}; e)$ is the extrinsic information about the set of code bits $\mathbf{C}$ of the SISO detector, which is fed back to the outer decoder as the intrinsic information of its coded bits. Before applying to the outer decoder, the extrinsic information is reordered to compensate for the pseudorandom interleaving introduced in the turbo encoder, yielding

$$\Lambda^o(\mathbf{C}; i) = \Pi^{-1}\left\{\Lambda^i(\mathbf{C}; e)\right\} \qquad (5.21)$$

2. The n_t outer SISO modules, in turn, process the soft information $\Lambda^o(\mathbf{c}_j(l); i)$, and compute refined estimates of soft information on both code bits $\mathbf{c}_j(l)$ and information bits $\mathbf{b}_j(l)$, based on the trellis structure of the channel codes, which is the channel code constraint. The computation proceeds as follows:
   - Compute the *a posteriori* information for information and code bits as follows:

$$\Lambda^o(b_j(l); p) = \log \frac{P\left\{b_j(l) = +1 | \Lambda^o(\mathbf{C}; i), \Lambda^o(\mathbf{B}; i), \text{decoding}\right\}}{P\left\{b_j(l) = -1 | \Lambda^o(\mathbf{C}; i), \Lambda^o(\mathbf{B}; i), \text{decoding}\right\}}, \forall j, l$$

$$(5.22)$$

$$\Lambda^o(c_j(l); p) = \log \frac{P\left\{c_j(l) = +1 | \Lambda^o(\mathbf{C}; i), \Lambda^o(\mathbf{B}; i), \text{decoding}\right\}}{P\left\{c_j(l) = -1 | \Lambda^o(\mathbf{C}; i), \Lambda^o(\mathbf{B}; i), \text{decoding}\right\}}, \forall j, l$$

$$(5.23)$$

The input $\Lambda^o(\mathbf{B}, i)$ is always initialized to zero, assuming equally likely source information bits.

- Compute the extrinsic information for information and code bits as follows:

$$\Lambda^o(\mathbf{B}; e) = \Lambda^o(\mathbf{B}; p) - \Lambda^o(\mathbf{B}; i) \qquad (5.24)$$

$$\Lambda^o(\mathbf{C}; e) = \Lambda^o(\mathbf{C}; p) - \Lambda^o(\mathbf{C}; i) \qquad (5.25)$$

The output, that is, the extrinsic information of the n_t outer decoders, provides intrinsic information to the inner SISO detector module after reordering to compensate for the random interleaving; thus

$$\Lambda^i(\mathbf{C}; i) = \Pi \left\{ \Lambda^o(\mathbf{C}; e) \right\} \qquad (5.26)$$

Steps 1 and 2 are repeated until the decoding algorithm converges.

3. Finally, an estimate of the message bits $\mathbf{B}$ is obtained by hard limiting the LLR $\Lambda^o(\mathbf{B}; p)$ at the output of the outer decoder, as shown by

$$\hat{\mathbf{B}} = \operatorname{sgn} \left\{ \Lambda^o(\mathbf{B}; p) \right\} \qquad (5.27)$$

Note that the outer decoder of the iterative decoding algorithm is made up of n_t parallel SISO channel decoders, each implemented with the generalized BCJR algorithm. A detailed explanation of the generalized BCJR algorithm was presented in Chapter 4.

5.6 DESIGN AND PERFORMANCE OF SISO DETECTORS

An important issue is the criterion used to optimize the inner SISO detector module in the iterative decoders. In this section, we consider the design of the inner SISO detector module using three possible criteria:

- Maximum *a posteriori* (MAP) probability estimation,
- Mean-squared-error minimization (MMSE), and
- Parallel soft interference cancellation (PSIC).

5.6.1 Performance Lower Bound

Consider a system that is equivalent to each of n_t transmitted signals received by a separate set of n_r antennas in such a way that each signal component is received with no interference from the other antennas. Each transmitted signal can be viewed as a $(1, n_r)$-BLAST signal with transmit power P/n_t, where P is the total transmitted power of the (n_t, n_r)- BLAST system. Let the ratio $P/\sigma^2 = \rho$ denote the total transmitted SNR. The overall received SNR of the kth information bit is

$$\bar{\rho} = \frac{\rho}{n_t} E\left\{\|\mathbf{h}_k\|^2\right\}$$
$$= \rho \frac{n_r}{n_t} \tag{5.28}$$

Equation (5.28) holds since the expectation $E\left\{\|\mathbf{h}_k\|^2\right\} = n_r$.

The (n_t, n_r) MIMO system may also be viewed as a $(1,1)$-BLAST system with average received SNR $= \bar{\rho}$. Hence, we may express the probability of error in terms of the average SNR per bit for a single transmit-receive antenna system. The bit error rate (BER) performance of a single transmit-receive antenna system with average received SNR $= \bar{\rho}$ for uncoded BPSK in a fixed and known fading environment is given by [106]

$$\mathrm{BER} = Q(\sqrt{2\bar{\rho}}) \tag{5.29}$$

where Q is the Gaussian Q-function.[1]

Since we use a soft decoding scheme in the receiver, we have an additional coding gain of $10\log_{10}(R_c d_{\min})$ in calculating the lower bound for the BER, where R_c and $d_{\min}$ are the code rate and free distance of the FEC, respectively. For convolutional codes, d_{free} will be used. Equation (5.29) is a lower bound on the BER performance of a T-BLAST scheme.

5.6.2 Detector Based on MAP Probability Estimation

To derive a simplified optimum receiver, we maximize the *a posteriori* metric of individual substreams separately. The received signal $\mathbf{y}(i) \in \mathbb{C}^{n_r \times 1}$ at the receive array at discrete time i is given by

$$\mathbf{y}(i) = \mathbf{H}(i)\mathbf{x}(i) + \mathbf{v}(i) \tag{5.30}$$

where $\mathbf{H}(i) \in \mathbb{C}^{n_r \times n_t}$, $\mathbf{x}(i) \in \mathbb{C}^{n_t \times 1}$, and $\mathbf{v}(i) \in \mathbb{C}^{n_r \times 1}$. Let $x_k(i)$ be the desired signal; we expect (5.30) as

$$\mathbf{y}(i) = \mathbf{h}_k(i)x_k(i) + \mathbf{H}_k(i)\mathbf{x}_k(i) + \mathbf{v}(i) \tag{5.31}$$

where $\mathbf{H}_k(i) = [\mathbf{h}_1(i), \mathbf{h}_2(i), \ldots, \mathbf{h}_{k-1}(i), \mathbf{h}_{k+1}(i), \ldots, \mathbf{h}_n(i)] \in \mathbb{C}^{n_r \times n_t - 1}$. The decision statistic of the kth substream using matched filtering is defined by the inner product $\mathbf{h}_k^\dagger \mathbf{y}_k(i)$, written as

$$y_k^1(i) = \underbrace{\mathbf{h}_k^\dagger \mathbf{h}_k x_k(i)}_{d_k} + \underbrace{\mathbf{h}_k^\dagger \mathbf{H}_k \mathbf{x}_k(i)}_{u_k} + \underbrace{\mathbf{h}_k^\dagger \mathbf{v}(i)}_{\bar{v}_k} \tag{5.32}$$

The three components d_k, u_k, and $\bar{v}_k$ identified in (5.32) are the desired signals obtained by the linear beamformer, the CAI, and the phase-rotated noise, respectively.

---

[1]The Gaussian function is defined by $Q(x) = \int_x^\infty \frac{1}{\sqrt{2\pi}} \exp\left(\frac{-t^2}{2}\right) dt$.

The soft output of the MAP detector for the kth substream is given by

$$\hat{x}_k = \arg \max_{x_k} P(x_k | y_k) \tag{5.33}$$

For brevity, in what follows we omit the sampling time index (i) in the equations. Let the interference of the kth substream be denoted by $\mathbf{u}_k \in U_k$, where

$$U_k = \left\{ (x_1, x_2, \ldots, x_{k-1}, 0, x_{k+1}, \ldots, x_{n_t}) : x_j \in \{+1, -1\}, j \neq k \right\}$$

is a set that spans $2^{n_t - 1}$ dimensions, and $\mathbf{u}_k$ is any $(n_t - 1)$-dimensional interference vector of the kth substream. The *a posteriori* probability of the kth substream is defined by

$$P(x_k | y_k^1) = P(y_k^1, x_k) / P(y_k^1) \tag{5.34}$$

where

$$P(y_k^1, x_k) = \sum_{\mathbf{u}_k \in U_k} P(y_k^1, x_k, \mathbf{u}_k) \tag{5.35}$$

and

$$P(y_k^1, x_k, \mathbf{u}_k) = P(y_k^1 | x_k, \mathbf{u}_k) P(x_k, \mathbf{u}_k)$$
$$= P(y_k^1 | x_k, \mathbf{u}_k) \prod_{j=1}^{n_t} P(x_j) \tag{5.36}$$

Here the probability distribution of the kth substream is formed by averaging out the contributions of interfering substreams. In the kth substream $(n_t - 1)$ interferences may be present, and therefore there are $2^{n_t - 1}$ possible interference patterns. In order to achieve the individual MAP decisions, we need to know the *a priori* probability of the interfering substreams $P(x_j), \forall j$. We use n_t-parallel SISO decoders to provide the *a priori* probabilities of the interfering substreams to the MAP detectors.

The corresponding LLR for the MAP SISO detectors are formed as sums rather than products of independent metrics. We define the log-likelihood metric as

$$\Lambda(x_k | y_k^1) = \Lambda(y_k^1 | x_k) + \Lambda(x_k) \tag{5.37}$$

where

$$\Lambda(x_k) = \log \frac{P(x_k = +1)}{P(x_k = -1)} \tag{5.38}$$

and

$$\Lambda(y_k^1 | x_k) = \log \frac{P(y_k^1 | x_k = +1)}{P(y_k^1 | x_k = -1)}$$
$$= \log \frac{\sum_{\mathbf{u}_k \in U_k} P(y_k^1 | x_k = +1, u_k) P(\mathbf{u}_k)}{\sum_{\mathbf{u}_k \in U_k} P(y_k^1 | x_k = -1, u_k) P(\mathbf{u}_k)}$$

$$= \log \frac{\sum_{\mathbf{u}_k \in U_k} \exp\left[-\frac{1}{2\sigma^2}(y_k^l - d_k - u_k)^2\right] \prod_{j \neq k} P(x_j)}{\sum_{\mathbf{u}_k \in U_k} \exp\left[-\frac{1}{2\sigma^2}(y_k^l + d_k - u_k)^2\right] \prod_{j \neq k} P(x_j)}$$

$$= \Lambda_{x_k}(x_k) + \Lambda_{\mathbf{u}_k}(x_k) \tag{5.39}$$

where

$$\Lambda_{x_k}(x_k) = \frac{2d_k y_k^l}{\sigma^2} \tag{5.40}$$

and

$$\Lambda_{\mathbf{u}_k}(x_k) = \log \frac{\sum_{\mathbf{u}_k \in U_k} P(\mathbf{u}_k) \exp\left(\frac{-1}{2\sigma^2}(u_k^2 - 2u_k(y_k^l - d_k))\right)}{\sum_{\mathbf{u}_k \in U_k} P(\mathbf{u}_k) \exp\left(\frac{-1}{2\sigma^2}(u_k^2 - 2u_k(y_k + d_k))\right)} \tag{5.41}$$

However, the computational complexity of the MAP detector is exponential in terms of the number of transmit antennas. One way of reducing the complexity is to use a suboptimal interference cancellation scheme, as described next.

5.6.3 Parallel Soft Interference Cancellation Receivers

A suboptimal solution may be found by using the max-log principle, which was described in Chapter 4. Specifically, the logarithm of the sum of exponentials in the numerator and the denominator of (5.39) can be approximated by the maximum of the exponents when the maximum of the exponents is much higher than the other exponent terms; we may then use (5.35) to (5.41) to write

$$\Lambda(y_k^l | x_k) \approx \frac{2d_k y_k^l}{\sigma^2} + \log\left(\frac{\max_{\mathbf{u}_k \in U_k} P(\mathbf{u}_k) \exp\left(\frac{-1}{2\sigma^2}(u_k^2 - 2u_k(y_k^l - d_k))\right)}{\max_{\mathbf{u}_k \in U_k} P(\mathbf{u}_k) \exp\left(\frac{-1}{2\sigma^2}(u_k^2 - 2u_k(y_k^l + d_k))\right)}\right) \tag{5.42}$$

In the second term of (5.42), the numerator and the denominator are Gaussian distributions centered at $(u_k + d_k)$ and $(u_k - d_k)$, respectively. In solving the corresponding optimization problems, we may therefore obtain different solutions for the interference vector $\mathbf{u}_k$, which is clearly undesirable. However, if we assume that the bit estimates from the previous iterations are highly reliable for all substreams, then for each substream there exists a unique interference pattern $\mathbf{u}_k$ such that $P(\mathbf{u}_k) \approx 1$. Thus, we may further approximate (5.42) as

$$\Lambda(y_k^l | x_k) \approx \underbrace{\frac{2d_k y_k^l}{\sigma^2}}_{\Lambda_{x_k}(x_k)} + \underbrace{\log\left(\frac{P(\mathbf{u}_k) \exp\left(\frac{-1}{2\sigma^2}(u_k^2 - 2u_k(y_k^l - d_k))\right)}{P(\mathbf{u}_k) \exp\left(\frac{-1}{2\sigma^2}(u_k^2 - 2u_k(y_k^l + d_k))\right)}\right)}_{\Lambda_{\mathbf{u}_k}(x_k)} \tag{5.43}$$

$$= \frac{2d_k y_k^l}{\sigma^2} - \frac{2}{\sigma^2}d_k(u_k) \tag{5.44}$$

Since we have expectations only of the interfering symbols, we may replace the exact information on the interfering substreams by the expectation of the interferences, as shown by

$$\Lambda_{\mathbf{u}_k}(x_k) = -\frac{2}{\sigma^2} d_k E\{u_k\} \tag{5.45}$$

where

$$E\{u_k\} = \sum_{j \neq k} \mathbf{h}_k^\dagger \mathbf{h}_j E\{x_j\} \tag{5.46}$$

SISO Decoders   We use n_t-parallel SISO decoders to provide the *a priori* probabilities of the transmitted substreams; for details of SISO decoders; see Section 4.3 of Chapter 4. Consider then the LLR of symbol x_j provided by the SISO decoder $\Lambda(x_j)$. Since, by definition,

$$\Lambda(x_j) = \log\left(\frac{P(x_j = +1)}{P(x_j = -1)}\right),$$

we may equivalently write

$$\frac{P(x_j = +1)}{P(x_j = -1)} = \exp(\Lambda(x_j)) \tag{5.47}$$

Furthermore, by using the fact that $P(x_j = +1) + P(x_j = -1) = 1$, we get the following component-wise relations:

$$P(x_j = +1) = \frac{\exp(\Lambda(x_j))}{1 + \exp(\Lambda(x_j))} \tag{5.48}$$

and

$$P(x_j = -1) = \frac{1}{1 + \exp(\Lambda(x_j))} \tag{5.49}$$

Hence, we may express the expectations $E\{x_j\}$ as a hyperbolic tangent function, as shown by

$$E\{x_j\} = \frac{(+1)\exp(\Lambda(x_j))}{1 + \exp(\Lambda(x_j))} + \frac{(-1)}{1 + \exp(\Lambda(x_j))}$$
$$\tag{5.50}$$
$$= \tanh(\Lambda(x_j)/2), \quad j = 1, 2, \ldots, n_t$$

Accordingly, the MAP formulation reduces to a soft interference cancellation scheme described by

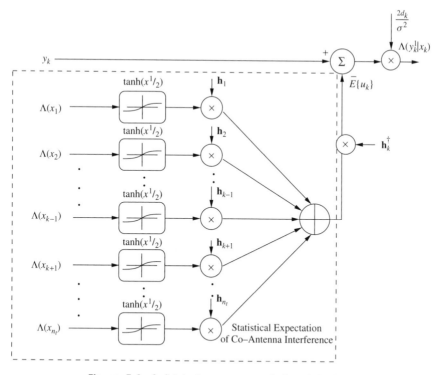

Figure 5.6 Soft interference cancellation detector.

$$\Lambda(y_k^1|x_k) = \frac{2d_k}{\sigma^2}\left[y_k^1 - \sum_{j\neq k}\mathbf{h}_k^\dagger\mathbf{h}_j \tanh(\Lambda(x_j)/2)\right] \qquad (5.51)$$

The resulting soft interference cancellation receiver is shown in Figure 5.6.

Experiment 1: Performance of the Map-Based Receiver In this experiment, we compare the performance of the MAP-based iterative receiver to that of the suboptimal soft interference canceling receiver. Figure 5.7 presents the results of computer simulations performed on four different receiver configurations:

- (1,1): system for an AWGN channel
- (8,12): coded V-BLAST system that relies on hard decisions in the receiver
- (8,12): T-BLAST with a MAP-based SISO detector
- (8,12): T-BLAST suboptimal soft interference cancellation detector

The transmitted power is maintained constant in all four configurations. The first receiver is noise-limited, whereas the other three BLAST configurations are called co-antenna interference- and fading-limited. The burst length L is 100 symbols,

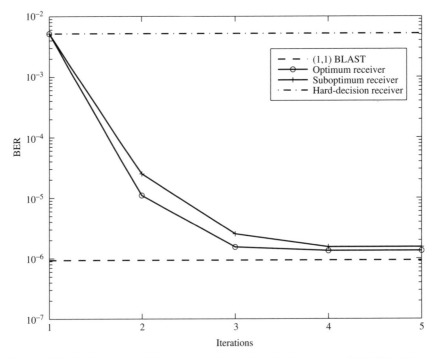

Figure 5.7 Performance of the proposed receivers for the encoded BLAST system.

20 of which are used for training. Each of the eight substreams utilizes rate-1/2 convolutionally encoded BPSK. The code generator used is (5,7) and eight different pseudo-random interleavers are used.

In Figure 5.7, the BER is plotted versus the number of iterations at SNR = 7dB. From the figure, we observe the following:

- The single antenna and V-BLAST configurations that rely on hard decisions are both naturally independent of the number of iterations.
- In direct contrast, the performances of the two T-BLAST configurations (reliant on soft decisions) improve with increasing number of iterations, approaching that of the noise-limited single-antenna system in about five iterations.
- The performance of the suboptimal T-BLAST configuration is close to that of the optimum one.

It is noteworthy that even with MAP detectors used as the inner decoders, iterative decoding can still provide a performance gain. The reason for performance improvement is that the MAP-based inner decoders are still suboptimal compared to the global MAP solution; the iterative decoders are used as a practical approximation to the global MAP receiver. Moreover, the turbo iterative process applies equally well to the suboptimal inner decoders; a soft interference cancellation inner decoder has a performance close to that of the MAP-based inner decoder, which is the reason for designing the simplified inner decoders.

5.6.4 Parallel Soft Interference Cancellation with Bootstrapping Channel Estimates

To extract the desired signal, we may use the maximum-ratio-combining (MRC) scheme since it is simple, yet effective; the MRC acts as a spatial match filter. In MRC, the beamforming weight vector is simply the estimated channel vector. To simplify the presentation, we assume that we have exact channel estimates. The decision statistic for the kth substream at the ith sampling interval obtained by the MRC is thus defined by

$$y_k^1(i) = \underbrace{\mathbf{h}_k^\dagger \mathbf{h}_k x_k(i)}_{d_k} + \underbrace{\sum_{j \neq k} \mathbf{h}_k^\dagger \mathbf{h}_j x_j(i)}_{u_k} + \underbrace{\mathbf{h}_k^\dagger \mathbf{v}(i)}_{\bar{v}_k}$$

$$= d_k(i) + u_k(i) + \bar{v}_k(i) \tag{5.52}$$

The terms designated as d_k, u_k and $\bar{v}_k$ in (5.52) are the desired signal obtained by the MRC, the CAI interference, and the phase-rotated noise, respectively. The receive-diversity gain achieved by MRC is $(n_r - n_t + 1)$, since $n_t - 1$ interferers are present. The MRC is known to be the optimum linear filter to combat multipath fading, provided that there is no co-channel interference. Unfortunately, in a correlated channel environment exemplified by BLAST, MRC is an inefficient way to extract the desired signal. In order to exploit the optimal behavior of MRC, we use a robust nonlinear interference canceler in front of the MRC, as described next.

Iterative Implementation of the Parallel Soft Interference Canceler
The soft-interference cancellation detector stage embodies three learning schemes:

1. An iterative soft interference cancellation scheme, which makes use of the implicit form of supervision provided by the FEC code via an iterative scheme.
2. A spatial channel matched filter, which is based on maximum ratio combining of the channel outputs since it is simple and effective. (For schemes (1) and (2), we use a short training sequence to estimate the channel matrix.)
3. Reestimation of the channel matrix, which uses the training sequence provided by the complete set of decoded information at each iteration.

In the iterative parallel soft interference canceler (IPSIC), the CAI is removed from the received signal before the receiver performs MRC to extract each desired signal. After canceling the interferers, the MRC exploits knowledge of the channel matrix to achieve n_r-fold diversity reception of each of the transmitted signals. To elaborate on this point, assume that we have a preliminary symbol estimate of each substream denoted by $\{\hat{x}_j\}$, $\quad j = 1, 2, \ldots, n_t$. Here again, to simplify the presentation, we omit the sampling index (i). The interference-free received vector pertaining to the kth substream is defined by

$$\mathbf{y}_k = \mathbf{y} - \sum_{j \neq k} \mathbf{h}_j \hat{x}_j \tag{5.53}$$

The decision statistic of the kth substream obtained by performing the MRC on the interference-free received vector is given by subtracting the inner product of $\mathbf{h}_k$ and $\sum_{j \neq k} \mathbf{h}_j \hat{x}_j$ from y'_k of (5.52), obtaining

$$x_k^1 = \underbrace{\mathbf{h}_k^\dagger \mathbf{h}_k x_k}_{d_k} + \underbrace{\sum_{j \neq k} \mathbf{h}_k^\dagger \mathbf{h}_j (x_j - \hat{x}_j)}_{\delta_k} + \underbrace{\mathbf{h}_k^\dagger \mathbf{v}}_{\bar{v}_k}$$

$$= d_k + \delta_k + \bar{v}_k \tag{5.54}$$

where we have introduced the new term

$$\delta_k = \sum_{j \neq k} \mathbf{h}_k^\dagger \mathbf{h}_j (x_j - \hat{x}_j).$$

If the receiver has exact knowledge of the interference, then δ_k reduces to zero and x_k^1 correspondingly reduces to

$$x_k^1 = \underbrace{\mathbf{h}_k^\dagger \mathbf{h}_k x_k}_{d_k} + \underbrace{\mathbf{h}_k^\dagger \mathbf{v}}_{\bar{v}_k} = \|\mathbf{h}_k\|^2 x_k + \bar{v}_k \tag{5.55}$$

where the squared Euclidean norm $\|\mathbf{h}_k\|^2$ is a chi-squared variate with $2n_r$ degrees of freedom. We normalize the channel such that $E\{|H_{kl}|^2\} = 1$ and thereby achieve n_r-fold receive diversity for each transmitted signal. However, since we do not know the actual symbol estimates of the transmitted substreams, we replace the exact symbol estimates $\{\hat{x}_j\}$ by their expectations $E\{x_j\}$. In particular, we use the *a priori* probabilities of the substreams to estimate the expectation $E\{x_j\}$, as shown by

$$E\{x_j\} = \sum_{x_j \in \{+1, -1\}} x_j P(x_j), \quad j = 1, 2, \ldots, n_t \tag{5.56}$$

Accordingly, we rewrite (5.54) as

$$x_k^1 = \underbrace{\mathbf{h}_k^\dagger \mathbf{h}_k x_k}_{d_k} + \underbrace{\sum_{j \neq k} \mathbf{h}_k^\dagger \mathbf{h}_j (x_j - E\{x_j\})}_{\hat{\delta}_k} + \underbrace{\mathbf{h}_k^\dagger \mathbf{v}(i)}_{\bar{v}_k}$$

$$= d_k + \hat{\delta}_k + \bar{v}_k \tag{5.57}$$

where $\hat{\delta}_k$ distinguishes itself from δ_k by replacing $\hat{x}_j$ with $E\{x_j\}$. Using (5.50), we may now express the expectation of the interference estimate as

$$
\begin{aligned}
E\{u_k\} &= \sum_{j \neq k} \mathbf{h}_k^\dagger \mathbf{h}_j \, E\{x_j\} \\
&= \sum_{j \neq k} \mathbf{h}_k^\dagger \mathbf{h}_j \, \tanh(\Lambda(x_j)/2)
\end{aligned}
\tag{5.58}
$$

Figure 5.8 shows the resulting structure of the soft interference cancellation scheme.

In order to use the interference cancellation scheme, we need to estimate the channel matrix. During the first decoding stage in the receiver, we use a short training sequence to obtain an initial estimate of the channel matrix. When performing linear beamforming and parallel soft interference cancellation, we benefit from a good estimate of the channel matrix. However, with a short training sequence, it may be difficult to achieve a good estimate for a slowly time-varying channel. To improve the channel estimate (within the packet) obtained as a result of the training, we reestimate the channel matrix using all estimated symbols of the packet at each subsequent iteration. This newly estimated matrix channel is used by the decoder to estimate spatial matched filter weights and interferences. The bootstrapping technique described herein is performed in a manner that tends to squeeze maximum information out of each packet.

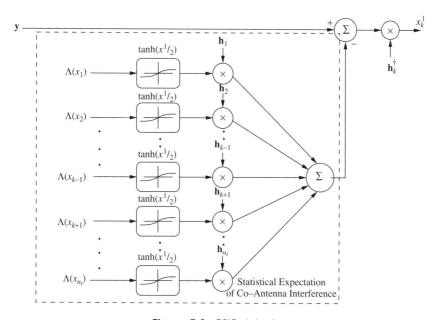

Figure 5.8 PSIC detector.

A serious limitation of the soft interference canceler is that it suffers from large error floors in a correlated environment. To avoid this error floor problem, we propose an MMSE-based decoder, which is described next.

5.6.5 MMSE Receiver

Recall the received signal $\mathbf{y}(i) \in \mathbb{C}^{n_r \times 1}$ at the receive array at time i defined in (5.30):

$$\mathbf{y}(i) = \mathbf{H}(i)\mathbf{x}(i) + \mathbf{v}(i)$$

where $\mathbf{H}(i) \in \mathbb{C}^{n_r \times n_t}$, $\mathbf{x}(i) \in \mathbb{C}^{n_t \times 1}$, and $\mathbf{v}(i) = \in \mathbb{C}^{n_r \times 1}$. Let $x_k(i)$ be the desired signal, yielding

$$\mathbf{y}(i) = \mathbf{h}_k(i)x_k(i) + \mathbf{H}_k(i)\mathbf{x}_k(i) + \mathbf{v}(i) \tag{5.59}$$

where $\mathbf{H}_k(i) = [\mathbf{h}_1(i), \mathbf{h}_2(i), \ldots, \mathbf{h}_{k-1}(i), \mathbf{h}_{k+1}(i), \ldots, \mathbf{h}_n(i)] \in \mathbb{C}^{n_r \times n_t - 1}$. Correspondingly, the decision statistic of the kth substream using a linear filter $\mathbf{w}_k$ is defined in (5.32), also reproduced here for convenient of presentation:

$$y_k^1(i) = \underbrace{\mathbf{w}_k^\dagger \mathbf{h}_k x_k(i)}_{d_k} + \underbrace{\mathbf{w}_k^\dagger \mathbf{H}_k \mathbf{x}_k(i)}_{u_k} + \underbrace{\mathbf{w}_k^\dagger \mathbf{v}(i)}_{\bar{v}_k}$$

where d_k, u_k, and $\bar{v}_k$ are the desired signals obtained by the linear beamformer, the CAI, and the CAI phase-rotated noise, respectively.

This time, we propose a multisubstream detector based on the MMSE principle and soft interference cancellation, which optimizes the interference estimate and the weights of the linear detector jointly in a manner similar to that of the multiuser receivers proposed in [162] and [40]. Specifically, we remove CAI from the linear beamformer output $y_k = \mathbf{w}_k^\dagger \mathbf{y}$ by writing

$$x_k^1 = \mathbf{w}_k^\dagger \mathbf{y} - u_k \tag{5.60}$$

where u_k (i.e., the CAI) is a linear combination of interfering substreams: u_k is a function of $\hat{\mathbf{x}}_k$, and x_k^1 is an improved estimate of the transmitted symbol x_k. For brevity, we have again omitted the sampling index (i). The performance of the estimator is measured by the instantaneous error $e_k = x_k - x_k^1$. The weights $\mathbf{w}_k \in \mathbb{C}^{n_t \times 1}$ and the interference estimate u_k are optimized by minimizing the mean-square value of the error between each substream and its estimate.

Problem 1 *Given the received signal $\mathbf{y}$ and the desired signal estimate x_k^1, find the weight vectors $\mathbf{w}_k$ and u_k by minimizing the cost function*

$$(\hat{\mathbf{w}}_k, \hat{u}_k) = \arg \min_{(\mathbf{w}_k, u_k)} E\left\{\|x_k - x_k^1\|^2\right\} \tag{5.61}$$

where the expectation E is taken over the noise and the statistics of the data sequence.

Solution 1 *The solution to Problem 1 is given by*

$$\hat{\mathbf{w}}_k = (\mathbf{P} + \mathbf{Q} + \Sigma_{n_r})^{-1} \mathbf{h}_k \tag{5.62}$$

$$\hat{u}_k = \mathbf{w}_k^\dagger \mathbf{z} \tag{5.63}$$

where

$$
\begin{aligned}
\mathbf{P} &= \mathbf{h}_k \mathbf{h}_k^\dagger && \in \mathbb{C}^{n_r} \\
\mathbf{Q} &= \mathbf{H}_k \left[\mathbf{I}_{(n_t-1)} - Diag(E\{\mathbf{x}_k\} E\{\mathbf{x}_k\}^\dagger) \right] \mathbf{H}_k^\dagger && \in \mathbb{C}^{n_r} \\
\Sigma_{n_r} &= \sigma^2 \mathbf{I}_{n_r}, \quad \sigma^2 > 0 && \in \mathbb{C}^{n_r} \\
\mathbf{z} &= \mathbf{H}_k \mathcal{E}[\mathbf{x}_k] && \in \mathbb{C}^{n_r \times 1}
\end{aligned}
$$

We used standard minimization techniques to solve the optimization problem formulated in (5.61) (see the appendix). In arriving at this solution, we used three assumptions:

$$E\left\{\mathbf{v}\mathbf{v}^T\right\} = \sigma^2 \mathbf{I}_{n_r}; \tag{5.64}$$

$$E\left\{\mathbf{x}\mathbf{v}\right\} = \mathbf{0}; \tag{5.65}$$

$$E\left\{x_i x_j\right\} = E\left\{x_i\right\} E\left\{x_j\right\} \quad \forall i \neq j \tag{5.66}$$

These conditions are achieved by independent and different space interleaving and time interleaving applied at the transmitter.

- *For the first iteration, we assume that $E\{\mathbf{x}_k\} = \mathbf{0}$, in which case (5.60) reduces to the description of a linear MMSE receiver for substream k:*

$$x_k^1(i) = \mathbf{h}_k^\dagger (\mathbf{H}\mathbf{H}^\dagger + \sigma^2 \mathbf{I})^{-\dagger} \mathbf{y} \tag{5.67}$$

- *With an increasing number of iterations, we assume that in the limit, $E\{\mathbf{x}_k\} \to \mathbf{x}_k$, in which case (5.60) simplifies to a perfect interference canceler:*

$$
\begin{aligned}
x_k^1 &= \left[(\mathbf{h}_k^\dagger \mathbf{h}_k + \sigma^2 \mathbf{I})^{-1} \mathbf{h}_k^\dagger \right]^\dagger (\mathbf{y} - \mathbf{H}_k \mathbf{x}_k) \\
&= (\mathbf{h}_k^\dagger \mathbf{h}_k + \sigma^2)^{-1} \mathbf{h}_k^\dagger (\mathbf{y} - \mathbf{H}_k \mathbf{x}_k)
\end{aligned} \tag{5.68}
$$

Solution 2 *The MMSE solution to the weight vector $\mathbf{w}_k$ problem requires the inversion of $n_r \times n_r$ matrices. A suboptimal solution to Problem 1 is obtained by ignoring the matrix $\mathbf{Q}$. We may represent the covariance matrix of residual interferences as follows:*

$$
\begin{aligned}
x_k^1 &= \mathbf{h}_k^\dagger ((\mathbf{h}_k \mathbf{h}_k^\dagger + \sigma^2 \mathbf{I})^{-1})^\dagger (\mathbf{y} - \mathbf{H}_k E\{\mathbf{x}_k\}) \\
&= ((\mathbf{h}_k^\dagger \mathbf{h}_k + \sigma^2)^{-1} \mathbf{h}_k^\dagger (\mathbf{y} - \mathbf{H}_k E\{\mathbf{x}_k\})
\end{aligned} \tag{5.69}
$$

This solution requires scalar inversion only, and the complexity of estimating (5.69) is equal to the complexity of the solution in (5.68). Note that the matrix $\mathbf{Q}$ *should not be mixed up with Q-function.*

In practice, we need the channel matrix, which means that finding a good estimate of the channel matrix is critical in an MIMO system with large transmit and receive antennas. The use of soft interference cancellation suffers from large error floors when channel estimation errors are present. Instead, we propose a bootstrapping channel estimation procedure intended to avoid the error floor. During the first iteration of the receiver, we use a short training sequence to estimate an initial channel matrix. We reestimate the channel matrix using reliably estimated symbols of each packet at each subsequent iteration and used by the detector to estimate spatial matched filter weights and interferences. The reliably estimated symbols are found by setting a threshold on the output LLRs. If the LLRs of the symbols exceed the threshold, then we use the hard decision of those symbols to update the channel values.

To acquire the expectation values of interfering substreams, we use n_t-parallel SISO decoders to provide the intrinsic probabilities of the transmitted bit streams. The intrinsic probabilities are obtained from the decoder soft outputs of the previous iterations using the following two relationships:

$$P(\{c_j\} = +1) = \frac{\exp(\Lambda(\{c_j\}))}{1 + \exp(\Lambda(\mathrm{Re}\{c_j\}))} \tag{5.70}$$

$$P(\{c_j\} = -1) = \frac{1}{1 + \exp(\Lambda(\{c_j\}))} \tag{5.71}$$

where $\Lambda(\{c_j\})$ is the soft output (formalized as the LLR) of symbol $\{c_j\}$ provided by the SISO decoder. The expectations of the transmitted bits are defined by

$$\mathcal{E}[\{c_j\}] = \frac{(+1)\exp(\Lambda(\{c_j\}))}{1 + \exp(\Lambda(\{c_j\}))} + \frac{(-1)}{1 + \exp(\Lambda(\{c_j\}))}$$
$$= \tanh(\Lambda(\{c_j\})/2), \quad j = 1, 2, \ldots, n_t \tag{5.72}$$

5.7 SIMULATIONS ON T-BLAST

The BLAST scheme considered for the simulation presented herein is a (16, 16) system. The packet length is 120 symbols, 20 of which are training symbols. Each of the 16 substreams uses a rate-1/2 convolutionally encoded BPSK. The code generator used is (5,7). The space-time interleavers are chosen randomly, and no attempt is made to optimize their design.

In validating the T-BLAST performance, we used the following channel models:

- A quasi-static Rayleigh fading channel, as depicted in Figure 5.9: A matrix of independent Rayleigh fading coefficients is generated, and the fading

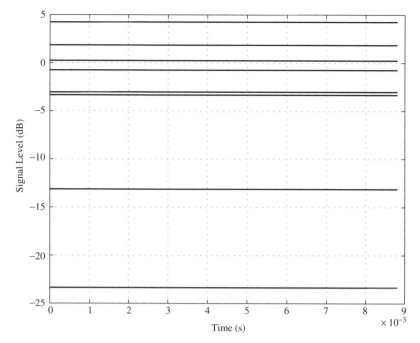

Figure 5.9 Quasi-static Rayleigh channel with eight transmit antennas.

coefficients are fixed over bursts of L symbols but are varied from one burst to the next.

- A slow fading Rayleigh fading channel, as depicted in Figure 5.10: Independent Rayleigh fading channels are generated according to the modified Jakes model [74] with maximum Doppler frequency between 0 and 30 Hz.

For each packet, a new realization of channel is chosen.

5.7.1 Performance of PSIC Receivers

In this experiment, we compare the performance of the PSIC receiver with and without bootstrapping channel reestimation. Computer simulations are performed on the following BLAST configurations with BPSK modulation:

- V-BLAST receiver using hard decisions,
- T-BLAST receiver using iterative soft interference cancellation (T-BLAST-MRC 1), and
- T-BLAST receiver using iterative soft interference cancellation with bootstrapping channel estimation (T-BLAST-MRC 2).

Experiment 2: Time-Invariant Channel   In this experiment, we consider independent channels that are time-invariant for the duration of a packet.

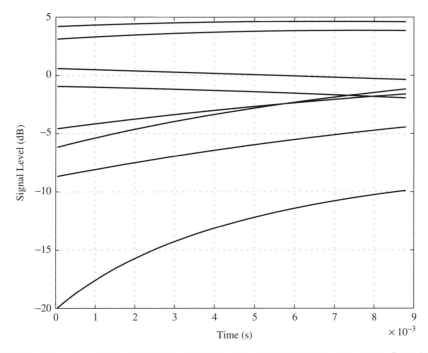

Figure 5.10 Slow-fading Rayleigh channel with eight transmit antennas; Doppler frequency = 10 Hz at a 1 GHz carrier frequency and a 10 km/H vehicle speed.

Figure 5.11 shows the BER versus the number of iterations for SNR = 9 dB. Naturally, the single-antenna system and the V-BLAST system are both independent of the number of iterations. In direct contrast, the performance of T-BLAST using both MRC 1 and MRC 2 improves with increasing number of iterations, with T-BLAST-MRC 2 performing slightly better than receiver T-BLAST-MRC 1. In particular, the performance of both T-BLAST receivers approaches that of the noise-limited single-antenna system (lower bound) in about four to five iterations.

In Figure 5.12, we show the BER performance versus SNR for V-BLAST and T-BLAST-MRC 2. The performance of T-BLAST-MRC 2 is shown for one, two, three, four, and seven iterations. As expected, the performance of both V-BLAST and T-BLAST improves with increasing SNR. The performance of T-BLAST approaches that of the single-antenna system (lower bound) in about four iterations. Moreover, the performance of T-BLAST exceeds that of V-BLAST in the course of two iterations.

Experiment 3: Slowly Time-Varying Channel Experiment 2 is repeated next with independent channels that vary slowly in time within a packet. The maximum Doppler frequency considered in the experiment is 20 Hz. Figures 5.13 and 5.14 show the BER versus number of iterations and the BER versus SNR

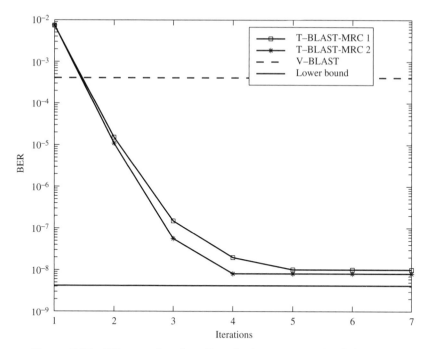

Figure 5.11 BER versus iterations; Doppler frequency = 0 Hz, SNR = 9 dB.

for one, two, four, and seven iterations for the T-BLAST-MRC receivers 1 and 2. From these figures, we make the following observations:

- The performances of both T-BLAST-MRC receivers improve with increasing number of iterations, with T-BLAST-MRC 2 outperforming T-BLAST-MRC 1. Both of these receivers outperform V-BLAST.

- After about four to five iterations, the T-BLAST-MRC receiver reaches a steady state but falls short of the single-antenna system (lower bound) in performance by a wider margin than in Experiment 2. This degradation in performance is attributed to the fact that a fixed matrix inadequately represents the time-varying channel matrix. For rapidly changing channels, the performance may be further decreased.

5.7.2 Performance of MMSE Receivers

In this section, we examine the performance of MMSE-based T-BLAST systems. Computer simulations are performed with the following BLAST configurations:

- Coded V-BLAST that relies on hard decisions,
- T-BLAST using the MMSE 1 solution expressed by (5.62), and
- T-BLAST using the MMSE 2 solution expressed by (5.69).

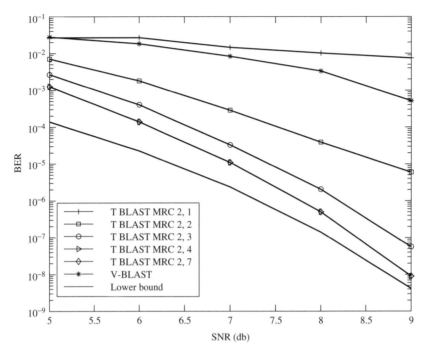

Figure 5.12 BER versus SNR; Doppler frequency = 0 Hz.

Experiment 4: Performance with Varying SNR Figure 5.15 shows the BER performance versus SNR for V-BLAST and T-BLAST-MMSE 1 and 2 for a 16 × 16 BLAST scheme. We observe the following:

- The performances of both V-BLAST and T-BLAST-MMSE improve with increasing SNR, which is to be expected.
- The performances of T-BLAST-MMSE receivers 1 and 2 improve with increasing number of iterations, with T-BLAST-MMSE 1 performing slightly better than T-BLAST-MMSE 2 for high SNR.
- The performance of T-BLAST exceeds that of V-BLAST within two iterations.

Experiment 5: Performance with an Increasing Number of Transmit Antennas Figure 5.16 shows the BER versus the number of transmit antennas for V-BLAST and T-BLAST-MMSE 2 for iterations 2 and 4 at SNR = 8 dB. The number of transmit antennas considered is 2, 4, 8, and 16. From the figure, we make the following observations:

- T-BLAST-MMSE 2 outperforms V-BLAST.
- The performance of T-BLAST receivers improves significantly with increasing number of transmit antennas. A significant performance increment is achieved

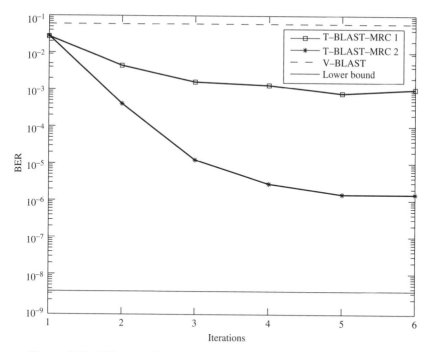

Figure 5.13 BER versus iterations; Doppler frequency = 20 Hz, SNR = 9 dB.

for the T-BLAST scheme with 16 transmit and 16 receive antennas compared to the T-BLAST scheme with 2 transmit and 2 receive antennas for the same operating conditions and the same total transmit power.

- The performance of V-BLAST increases from the 2×2 antenna scheme to the 8×8 antenna scheme. This improvement diminishes for the 16×16 antenna scheme due to the domination of the worst decoded layer in the average BER.

This example illustrates the robustness of the T-BLAST scheme in the presence of co-antenna interferences. T-BLAST performs better with large numbers of transmit antennas and achieves significant diversity and coding gain: the larger the interleaver depth, the higher the coding gain. Since the decoder outputs are used to estimate the soft interferences, the scheme is also guaranteed to achieve maximum receiver diversity. In direct contrast, the performance of V-BLAST decreases with larger numbers of transmit antennas due to the increased amount of co-channel interference. Since V-BLAST utilizes no transmit diversity, it has no additional means to compensate for the increased co-channel interference.

5.7.3 MMSE versus MRC for T-BLAST

Next, we compare the performance of MRC-based soft interference cancellation receivers with that of MMSE-based receivers. Computer simulations are performed on the following BLAST configurations with QPSK modulation:

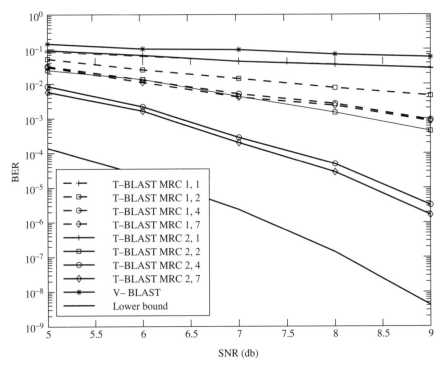

Figure 5.14 BER versus SNR; Doppler frequency = 20 Hz.

1. D-BLAST with no edge wastage and relying on hard decisions,
2. T-BLAST-MMSE 1, and
3. T-BLAST-MRC 2.

Experiment 6: Performance with a Time-Invariant Channel

Figure 5.17 shows BER performance versus SNR for D-BLAST and T-BLAST receivers for iterations 1, 2, and 5. As expected, the performance of both D-BLAST and T-BLAST improves with increasing SNR. The performance of both T-BLAST receivers improves with increasing iterations and exceeds that of D-BLAST within two iterations. A significant gain (7 dB) is achieved by the T-BLAST scheme over D-BLAST. T-BLAST-MMSE performs around 0.75 dB better than T-BLAST-MRC.

Experiment 7: Performance with a Time-Varying Channel

Experiment 6 is repeated with independent channels that vary slowly in time within a burst (packet). The maximum Doppler frequency considered here is 20 Hz. Figure 5.18 shows the BER performance after five iterations versus SNR for T-BLAST-MMSE and T-BLAST-MRC. In this figure we see that the performance

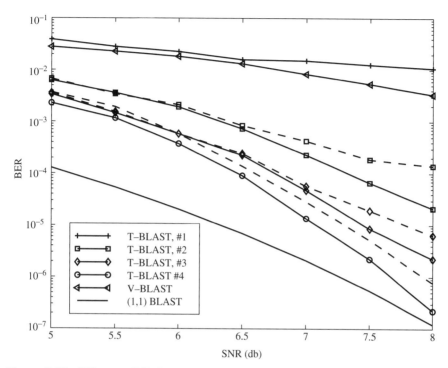

Figure 5.15 BER versus SNR. The continuous lines and dashed lines represent, respectively, the performance of T-BLAST-MMSE 1 and 2.

of both T-BLAST receivers improves with SNR, with T-BLAST-MMSE outperforming T-BLAST-MRC by a significant margin (2 dB). This result illustrates the robustness of the MMSE receiver for channel estimation errors. Moreover, a performance degradation of about 6 dB is observed by comparing Figure 5.18 to Figure 5.17 due to possible channel estimation errors for the time-varying channel.

5.7.4 Interleaver Dependence

Computer simulations have also been performed with the following configurations with QPSK modulation for the time-invariant channel:

- T-BLAST (random interleavers),
- T-BLAST with diagonal layering interleavers with no edge waste (T-D-BLAST), and
- Traditional D-BLAST but with no edge wastage.

Experiment 8: Interleaver Design Dependence Figure 5.19 shows the BER performance versus SNR for T-D-BLAST and T-BLAST for iterations one,

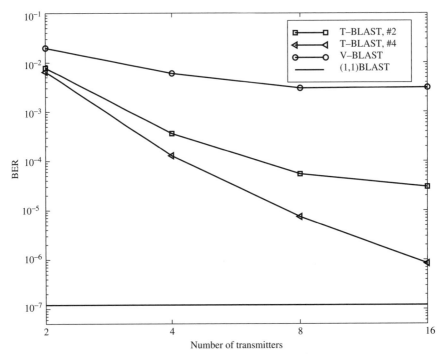

Figure 5.16 BER versus number of transmit antennas for T-BLAST-MMSE receiver 1; SNR = 8 dB.

two, four, and five; T-D-BLAST refers to D-BLAST using an iterative receiver. The figure also shows the performance of traditional D-BLAST. Based on this figure, we make the following observations:

- T-BLAST and T-D-BLAST outperform traditional D-BLAST within two iterations. In particular, at the 10^{-4} BER level, traditional D-BLAST performs 7 dB worse compared to T-D-BLAST with an iterative receiver.
- The performance of T-BLAST and T-D-BLAST is virtually identical.

The last observation indicates that random interleaving is sufficient for T-BLAST, since the independent subchannels are random and vary from packet to packet.

Experiment 9: Interleaver Size Dependence   In Figure 5.20, we show the BER performance at iteration 5 versus SNR for different sizes of interleavers. Note that the width of the interleaver equals the number of transmit antennas, which is 16 for the scenario considered herein. We observe again that the performances of T-BLAST and T-D-BLAST are practically identical.

The performance of both schemes improves with interleaver size. Note that interleaver sizes 100, 200, and 400 correspond to packet sizes 25, 50, and 100 of

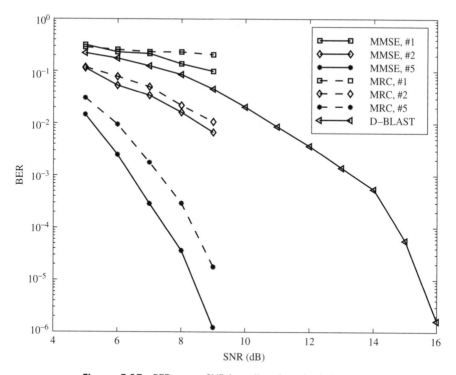

Figure 5.17 BER versus SNR for a time-invariant channel.

modulated signals. The packet size 100 is the scenario in wireless communications, achieving the best performance of all the sizes considered. In particular, at BER 10^{-5}, T-D-BLAST with an interleaver size of 400 gains 5 dB and 1.5 dB SNR over T-D-BLAST with interleaver sizes of 100 and 200, respectively. This gain in SNR can be attributed to many factors. One reason may be the possible but rare presence of bad interleavers as interleaver size increases.

Remark 5 *The iterative process may seem to increase the complexity of the entire receiver. However, with iterative detection, we can use suboptimal methods for signal detection and therefore reduce the overall complexity. In particular, T-BLAST has no optimal ordering step that contributes to the computational complexity (growing as the fourth power of n_t) of the V-BLAST scheme.*

5.7.5 Results Using Indoor Channel Measurements

This section compares the performance of QPSK-modulated T-BLAST with that of a corresponding horizontally coded V-BLAST using real-life indoor channel measurements with various MIMO configurations. The channel measurements were acquired using a narrowband test bed at Bell Labs of Lucent Technologies, Crawford Hill, New Jersey, during October 2000, in an indoor environment. At the

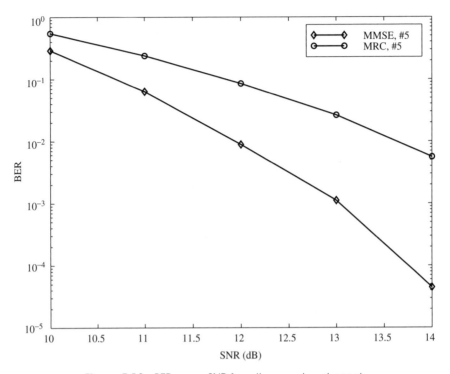

Figure 5.18 BER versus SNR for a time-varying channel.

transmit end, each substream of 100 information bits was independently encoded using a rate-1/2 convolutional code generator (5,7) and then interleaved using space-time interleavers. The space interleavers are designed using diagonal layering interleavers (Figure 5.2). The time interleavers were chosen randomly-, and no attempt was made to optimize their design.

The received signal was constructed based on knowledge of the measured channel characteristics, and the performance of T-BLAST was evaluated over a wide range of SNRs using various BLAST combinations. For the first two experiments, T-BLAST was evaluated with the exact channel matrix. In the third experiment, channel estimation was performed using a short training sequence followed by iterative channel estimation.

The performance of T-BLAST was compared with that of a corresponding horizontally coded V-BLAST in which each of the substreams was provided with an amount of channel coding equal to that used in T-BLAST. (Recall that V-BLAST does not use any space-time coding or iterative decoding.) The V-BLAST algorithm used here uses the following major provisions:

- Finding the optimal order of detection.
- Decoding the strongest signal using MMSE nulling vectors.

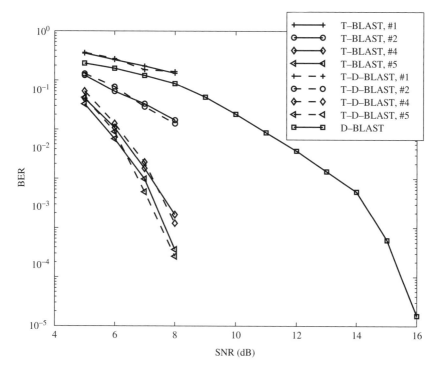

Figure 5.19 Performance comparison of D- and T-BLAST schemes.

- Decoding the substreams using the maximum *a posteriori*–based SISO channel decoders.
- Cancellation of interference due to the decoded signal using hard decisions.
- Finding and decoding the strongest signal component of the remaining signals, and so on.

Experiment 1: *T-BLAST versus V-BLAST,* $n_t = 5, 6, 7, 8$ *and* $n_r = 8$

Figure 5.21 compares the BER performance of T-BLAST (solid trace) and coded V-BLAST (broken trace) for antenna configurations of eight receive and five, six, seven, or eight transmit antennas. From the figure, we make the following observations:

- T-BLAST gives the best performance obtained within the first 10 iterations. The BER performance of both V-BLAST and T-BLAST improves with decreasing number of transmitters, with T-BLAST outperforming V-BLAST in all four cases.
- In terms of V-BLAST performance, a substantial gain in BER performance is realized with fewer transmit antennas. However, V-BLAST falls short of T-BLAST performance by a significant margin with more transmit antennas.

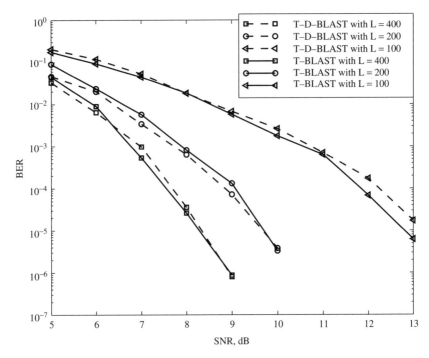

Figure 5.20 Performance variation with varying interleaver size L.

For example, T-BLAST achieves a 2–3 dB gain over V-BLAST for $n_t = 7$ and $n_r = 8$, whereas only a 0.5 dB gain is attained when $n_t = 5$ and $n_r = 8$.

Experiment 2: T-BLAST versus V-BLAST, $n_t = 8$ and $n_r = 5, 6, 7$ Consider next the case of BLAST configurations with fewer receive antennas than transmit antennas; Figure 5.22 compares the BER performance of T-BLAST (solid trace) with that of horizontally coded V-BLAST (broken trace):

- With antenna configurations of eight transmit antennas and five to eight receive antennas, T-BLAST gives the best performance within the first 10 iterations.
- The figure also reveals a major limitation of V-BLAST system: the inability to work efficiently with fewer receive antennas than transmit antennas.
- In T-BLAST, BER performance improves with increasing number of receivers, with T-BLAST outperforming V-BLAST in all four cases. Moreover, increasing the number of receivers from seven to eight offers a marginal benefit.

Experiment 3: T-BLAST versus V-BLAST, $n_t = n_r = 8$ and Iterative Channel Estimates   In Figures 5.23 and 5.24, we compare the performances

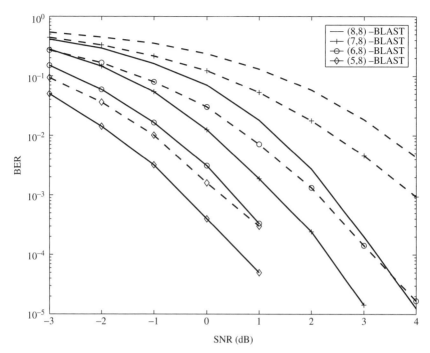

Figure 5.21 BER performance for $n_t = 5$, 6, 7, and 8 and $n_r = 8$, using convolutional code with rate $R = 1/2$ and constraint length 3 and QPSK modulation.

of (1) an iterative decoder with initial channel estimation using 16 training symbols and (2) an iterative decoder with initial channel estimation and iterative refined channel estimation. The BER performance is compared for T-BLAST architectures with perfect channel knowledge and with perfect channel and interference knowledge. Figure 5.23 shows the convergence of the iterative detection and decoding (IDD) receivers at SNR = 3 dB. Figure 5.24 shows the BER performance of IDD receivers versus SNR under various conditions considered at iterations 1 and 9. Although, the BER performance of the decoder with iterative channel estimation is initially (i.e., at the first iteration) worse than that of the decoder with perfect channel knowledge, at the fifth iteration it becomes very close to the performance of the decoder with perfect channel knowledge. Moreover, both decoders converge close to the decoder which has knowledge of both the channel and the interference. The BER performance of the decoder with initial channel estimates only is about 2–4 dB worse than that of the other schemes because of channel estimation errors.

5.7.6 Results with Correlated Channels (Indoor and Outdoor Measurements)

Theoretical investigations of correlated fading environments show that not only the i.i.d. MIMO links but also the spatially correlated MIMO links can offer a

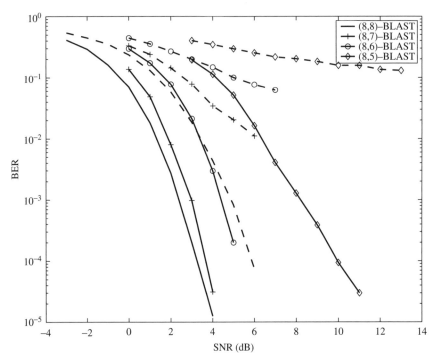

Figure 5.22 BER performance for $n_t = 8$ and $n_r = 5$, 6, 7, and 8, using convolutional code with rate $R = 1/2$ and constraint length 3 and QPSK modulation.

capacity that grows linearly with n but the growth rate is 10–20% smaller for the spatially correlated links [26, 138]. Moreover, the narrowband MIMO link measurements of fixed wireless systems outdoors show the possible presence of spatially correlated channels and significant variation in channel capacity limits due to spatial correlation [53]. In this section, we explore the suitability of the T-BLAST architecture in correlated Rayleigh fading MIMO environments. In particular, for an MIMO system with a large number of receive antennas, near-optimal performance can be achieved by the T-BLAST architecture in spatially and temporally correlated Rayleigh fading environments.

We provide the BER and capacity performances of T-BLAST architectures. The BER performances are provided for simulated quasi-static and time-varying correlated MIMO channels. For capacity analysis, the performance results were based on the measured real-life channels in both indoor and outdoor environments. For all the results at the transmit end, each substream was independently encoded using a rate-1/2 convolutional code generator (5,7) and then interleaved using space-time interleavers. The space interleavers were designed using diagonal layering interleavers [129]. The time interleavers were chosen randomly, but no attempt was made to optimize their design. In all the experiments presented next, it was assumed that the receiver had perfect knowledge of the channel matrix.

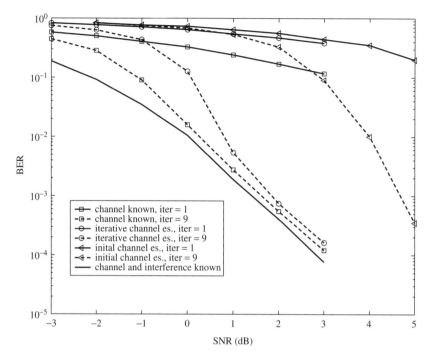

Figure 5.23 BER performance with iterative channel estimation for $n_t = 8$ and $n_r = 8$, using convolutional code with rate $R = 1/2$ and constraint length 3 and QPSK modulation.

The signals used were QPSK modulated in all the simulations. References [129] to [131] provide additional performance evaluations of the T-BLAST architecture.

Experiment 1: Performance with Spatially Correlated Quasi-Fading Channels for $n_t = n_r = 4$ This experiment demonstrates the performance of T-BLAST in the presence of fading signal correlations in a wireless environment. In particular, the experiment shows that by increasing the length of the random time interleavers, near-optimal performances in correlated fading environments is attained.

Random realizations of correlated Rayleigh fading channels were generated using the following covariance matrix structure:

$$E\left[\frac{\mathbf{H}^\dagger \mathbf{H}}{n_r}\right] = \begin{bmatrix} 1 & \gamma & \gamma & \gamma \\ \gamma & 1 & \gamma & \gamma \\ \gamma & \gamma & 1 & \gamma \\ \gamma & \gamma & \gamma & 1 \end{bmatrix} \tag{5.73}$$

Figure 5.25 shows the BER performances of T-BLAST for $\gamma = 0.3$, 0.7, and 0.9 for a packet size of 100 symbols. In all cases, the BER performance results

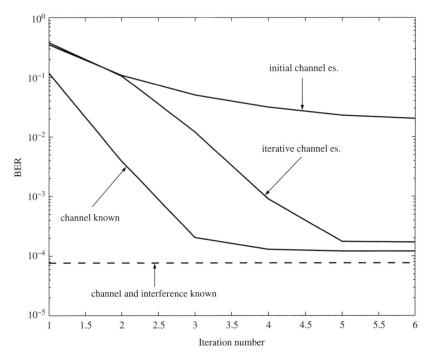

Figure 5.24 Convergence behaviors of IDD receivers under various conditions: BER performance with number of iterations for $n_t = 8$ and $n_r = 8$ using convolutional code with rate $R = 1/2$ and constraint length 3 and QPSK modulation.

are compared to those of T-BLAST with perfect knowledge of interference plotted with corresponding thick lines. As we see from the figure, all three cases converge close to the decoder which has knowledge of the interference. The shift in the BER performances between the plots is related to the channel capacity limit constraints rather than the limitations of T-BLAST. Moreover, the theoretical SNRs required to transmit 4 bits per channel through channels described by $\gamma = 0.3$, 0.7, and 0.9 are 0, 1.2, and 2.5 dB, respectively.

In Figure 5.26, we show the BER performance at iterations 2 and 5 versus SNR for different packet sizes for correlated Rayleigh fading channels generated with $\gamma = 0.7$ and 0.9. The performance of T-BLAST asymptotically (at higher SNRs) improves with packet size. Moreover, the performances of T-BLAST exceeds that of V-BLAST in two iterations at low SNR.

Experiment 2: Performance with Mobile Channels The time-varying channel is generated according to the Jakes model [74]. In a time-varying environment, we assume that the channel is perfectly tracked by the receiver.

Performance with Varying n_t and n_r for $M = 100$ This example demonstrates that T-BLAST achieves optimal performance (performance of a coded

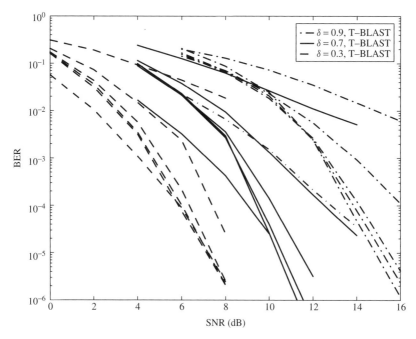

Figure 5.25 Performance of T-BLAST in spatially correlated Rayleigh fading environments $n_t = n_r = 4$, using convolutional code with rate $R = 1/2$ and constraint length 3 and QPSK modulation. Dashed-dot curves: $\gamma = 0.9$; Continuous curves: $\gamma = 0.7$; Dashed curves: $\gamma = 0.3$.

AWGN channel with total power P) in a Rayleigh fading environment when the number of transmit and receive antennas is large. In particular, we show the BER performance of coded V-BLAST and T-BLAST with increasing numbers of transmit and receive antennas for both quasi-static and time-varying environments.

Figure 5.27 shows the BER versus the number of transmit antennas for V-BLAST and T-BLAST for iterations 1, 2, and 8; $n_t = n_r = 2$, 4, 8, and 16 at SNR $= 3$ dB. From the figure, we see that T-BLAST outperforms V-BLAST in two iterations. The performance of T-BLAST receivers improves significantly with increasing number of transmit and receive antennas. A significant performance increment is achieved for T-BLAST with 16 transmit and 16 receive antennas compared to T-BLAST with 2 transmit and 2 receive antennas for the same operating conditions and the same total transmit power. Most important, for 16 transmit and 16 receive antennas, the performance of T-BLAST for the time-varying channel reaches the performance of the noise-limited AWGN channel. In a quasi-fading environment, the optimal performance is achieved for large n_r. The performance of V-BLAST increases from the 2×2 scheme to the 8×8 scheme. In fact, this improvement diminishes and a performance degradation is observed for the 16×16 antenna scheme compared to that of

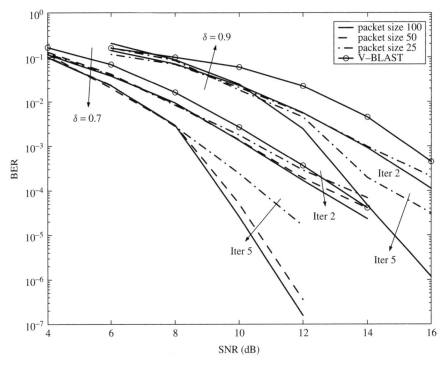

Figure 5.26 T-BLAST versus V-BLAST with varying interleaver sizes $n_t = n_r = 4$ in a spatially correlated Rayleigh fading correlated channel environment, using convolutional code with rate $R = 1/2$ and constraint length 3 and QPSK modulation. Continuous curves: packet size = 100; dashed curves: packet size = 50; dashed-dot curves: packet size = 25; continuous-circle curves: V-BLAST with packet size of 100.

the 8×8 antenna scheme. Note that the V-BLAST-OSIC algorithm is suitable only in a quasi-static environment. Thus, we do not show the performance of V-BLAST-OSIC for time-varying channel environments.

Moreover, this example illustrates the robustness of T-BLAST in the presence of co-antenna interferences. In direct contrast, the performance of the corresponding V-BLAST decreases with larger numbers of transmit antennas due to the increased amount of co-channel interference. Since V-BLAST does not utilize any transmit diversity, it has no additional means to compensate for the increased co-channel interference.

Performance with Varying Doppler Spreads and Interleaver Sizes for $n_t = n_r = 4$ This example demonstrates that by increasing the size of the random time interleavers, we can achieve better performance when high temporal channel correlations exist. Figure 5.28 shows the performance of T-BLAST for iterations 1 and 5 with different Doppler spreads and packet sizes. The performance improves significantly when temporal diversity is available. At BER $= 10^{-5}$, the performance

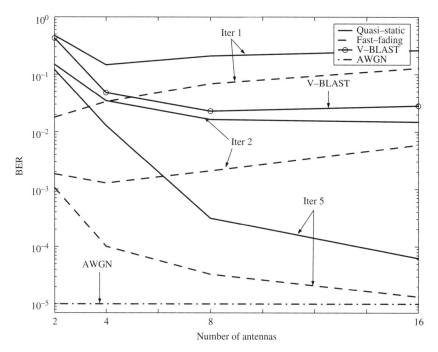

Figure 5.27 T-BLAST versus V-BLAST with varying numbers of transmit and receive antennas, $n_t = n_r = 2$, 4, 8, and 16, using convolutional code with rate $R = 1/2$ and constraint length 3 and QPSK modulation. Continuous curves: quasi-static fading channels; dashed curves: fast-fading channels; continuous-circle curve: V-BLAST; dashed-dot curve: lower bound on the performances.

corresponding to the Doppler frequency of 100 Hz has a gain of 4 dB over the Doppler frequency of 10 Hz and for packet size M = 100. For a larger packet size of M = 1000, the slow fading channel (Doppler frequency of 10 Hz) has a gain of 4 dB over the system with M = 100 and the same Doppler frequency. No significant gain is observed by increasing the packet size further for both cases. The performance gain observed with larger packet sizes in this case is due to the higher coding gain that comes from the better time selectivity obtained by using a larger packet size and interleaving (which reduces temporal correlations). The shift in the BER curves between the two different packet sizes is due to the coding gain. The optimal performances are about 2 dB away from the performance of the AWGN channel using the same channel code with equal total power at BER $= 10^{-5}$.

5.7.7 Spectral Efficiency Using Real-Life Data

Another measurement of performance is the information transmission rate of T-BLAST with the given channel code (convolutional code with rate $R = 1/2$ and constraint length 3), block size (100 symbols/transmission), and modulation

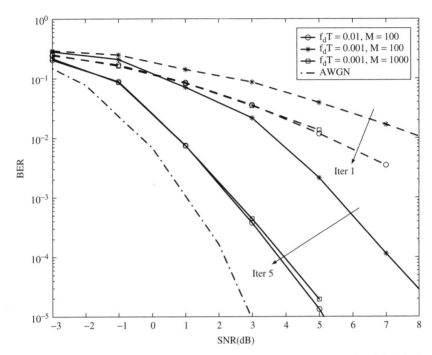

Figure 5.28 BER performance of T-BLAST in a temporarily correlated Rayleigh fading environment with various Doppler spreads and interleaver sizes for $n_t = 4$ and $n_r = 4$, using convolutional code with rate $R = 1/2$ and constraint length 3 and QPSK modulation. Dashed curves: performances at iteration 1; continuous curves: performances at iteration 5; dashed-dot curve: lower bound on the performances.

(QPSK). The information rates of T-BLAST and V-BLAST with the given channel code are evaluated as follows: We fix the target BER for error-free communication to be achieved at 10^{-3} and derive the SNR necessary to achieve this targeted BER for both V-BLAST and T-BLAST receivers. We consider both indoor (i.i.d. channels) and outdoor (correlated channels) in the analysis of spectral efficiency.

Indoor Channels Table 5.1 summarizes the SNR necessary to achieve this targeted BER 10^{-3}, information rate of T-BLAST, the actual capacity of the (8,8) matrix channel, and the percentage of channel capacity achieved by T-BLAST referenced to the capacity of the (8,8) matrix channel for the various antenna configurations considered herein. The corresponding values for V-BLAST are given in Table 5.2. In terms of T-BLAST performance, we observe the following results:

- A power gain of 0.5–4 dB is achieved over the V-BLAST system
- With T-BLAST, 88% of channel capacity is attained using the antenna configuration (8,8); the corresponding value is 60% for V-BLAST.

Table 5.1. Spectral efficiency of T-BLAST in an indoor environment

Configuration (n_t, n_r)	SNR ρ (dB)	Capacity of T-BLAST C_T Bits/Channel Use	Channel Capacity C Bits/Channel Use	Percentage Capacity
(5,8)	0.0	5	6.80	74 %
(6,8)	0.5	6	7.36	81 %
(7,8)	1.0	7	7.93	88 %
(8,8)	2.0	8	9.14	88 %
(8,7)	3.0	8	10.46	76 %
(8,6)	4.0	8	11.89	67 %

Table 5.2. Spectral efficiency of V-BLAST in an indoor environment

Configuration (n_t, n_r)	SNR ρ (dB)	Capacity of V-BLAST C_V Bits/Channel Use	Channel Capacity C Bits/Channel Use	Percentage Capacity
(5,8)	0.5	5	7.36	67 %
(6,8)	2.5	6	9.70	61 %
(7,8)	4.0	7	11.89	59 %
(8,8)	5.0	8	13.40	60 %

Moreover, the maximum possible information transmission rate of the T-BLAST and V-BLAST systems is 8 bits/channel use. At SNR $= 1.1$ dB, the channel capacity is 8 bit/channel use. Given this reference, T-BLAST and V-BLAST attain the Shannon theoretical capacity limit within 0.9 and 3.9 dB of average SNR, respectively.

It is informative to examine the covariance matrix of the channels considered. As a sample case, the estimated channel correlation matrix of the indoor channel for the (5,8) antenna configuration is

$$E\left[\frac{\mathbf{H}^\dagger\mathbf{H}}{n_r}\right] = \begin{bmatrix} 1.0000 & 0.5091 & 0.4099 & 0.0972 & 0.2804 \\ 0.5091 & 1.0000 & 0.1931 & 0.1464 & 0.1431 \\ 0.4099 & 0.1931 & 1.0000 & 0.4424 & 0.2075 \\ 0.0972 & 0.1464 & 0.4424 & 1.0000 & 0.0489 \\ 0.2804 & 0.1431 & 0.2075 & 0.0489 & 1.0000 \end{bmatrix} \tag{5.74}$$

where the off-diagonal components of the matrix are close to or smaller than 1/2.

The channel correlation matrix is the result of correlations between individual subchannels, which is somewhat controlled by the scattering phenomena occurring in the measurement site, the distance between individual antennas at both the transmit and receive ends, and the antenna polarizations.

Outdoor Channels   The experiments presented in the remainder of this section are based on the preliminary outdoor channel measurements acquired from Lucent

Technologies, Bell Laboratories at Crawford Hill, New Jersey, and were collected using a BLAST configuration with five transmit antennas and seven receive antennas. The distance between the locations of the transmitter and receiver was about 2 km. The system operated at a carrier frequency of 1.95 GHz with a 30 kHz transmission bandwidth.

Table 5.3 shows the SNR necessary to achieve a targeted BER of 10^{-3}, the information rate of T-BLAST in bits/channel use, the Shannon theoretical capacity of a (5,7) matrix channel, and the percentage of channel capacity achieved by T-BLAST referenced to the capacity of a (5,7) matrix channel for the antenna configurations considered in the experiments. The corresponding values of V-BLAST are listed in Table 5.4. In terms of T-BLAST performance, a power gain of 2–3.5 dB is achieved over V-BLAST performance. Moreover, T-BLAST attains 86% of the channel capacity with antenna configuration (5,7), whereas V-BLAST attains only 57% of the channel capacity with the same configuration.

Furthermore, the maximum possible information transmission rate of T-BLAST and V-BLAST is 5 bits/channel use. Meanwhile, at SNR = 0.2 dB, the channel capacity is 5 bits/channel use. Given this frame of reference, T-BLAST and V-BLAST attain the Shannon theoretical capacity limit within 1.3 and 4.8 dB of the average SNR, respectively. The estimated channel correlation matrix in the (5,7) case for the outdoor channel is given by

$$E\left[\frac{\mathbf{H}^{\dagger}\mathbf{H}}{n_r}\right] = \begin{bmatrix} 1.0000 & 0.8640 & 0.2157 & 0.2022 & 0.3329 \\ 0.8640 & 1.0000 & 0.3219 & 0.3097 & 0.5681 \\ 0.2157 & 0.3219 & 1.0000 & 0.5777 & 0.5638 \\ 0.2022 & 0.3097 & 0.5777 & 1.0000 & 0.6310 \\ 0.3329 & 0.5681 & 0.5638 & 0.6310 & 1.0000 \end{bmatrix} \qquad (5.75)$$

Table 5.3. Spectral efficiency of T-BLAST in an outdoor environment

Configuration (n_t, n_r)	SNR ρ (dB)	T-BLAST C_T Bits/Channel use	Channel C Bits/Channel use	Percentage Capacity
(5,7)	1.5	5	5.82	86 %
(4,7)	0.5	4	5.13	78 %
(3,7)	0.5	3	5.13	58 %

Table 5.4. Spectral efficiency of V-BLAST in an outdoor environment

Configuration (n_t, n_r)	SNR (dB)	V-BLAST C_V Bits/Channel Use	Channel C Bits/Channel Use	Percentage Capacity
(5,7)	5.0	5	8.73	57 %
(4,7)	3.0	4	6.91	58 %
(3,7)	2.5	3	6.54	43 %

The off-diagonal components of the correlation matrix in (5.75) are in general higher than those of the correlation matrix of indoor channels in (5.74), as expected.

5.8   SUMMARY AND DISCUSSION

In this chapter, we considered another BLAST architecture called T-BLAST. In T-BLAST, the data stream is split into parallel substreams and each substream is encoded independently using block codes. The encoded substreams are then space-time interleaved using independently generated random time interleavers and diagonal layering space interleavers. We refer to these codes as random layered space-time (RLST) codes. The structure of these RLST codes achieves two objectives: (1) an iterative turbo-like receiver for jointly decoding the simultaneously transmitted substreams with low complexity and (2) the realization of a significant percentage of the MIMO channel capacity in a computationally feasible manner. A hallmark of T-BLAST is that the error performance improves with the number of iterations of the decoding algorithm and, most importantly, it exceeds the performance of corresponding coded V-BLAST in two iterations. This low-complexity decoding is achieved by splitting the global maximum likelihood decoders into two stages of decoding and feeding extrinsic information from the output of one decoding stage to the input of the next decoding stage, which permits the iterative decoding process to take its natural course in response to the received noisy signal and channel code constraint.

We demonstrated the performance of T-BLAST using simulated and real-life wireless channel data with various antenna configurations, including fewer receive antennas than transmit antennas in an indoor environment. The iterative detection decoding receiver with iterative channel estimation improves the BER performance rapidly at each iteration, and it converges close to a decoder with knowledge of the channel and interference within four to five iterations. Moreover, we have shown that, by using real-life data at a target BER of 10^{-3}, a power gain of 2 to 4 dB is achieved over the corresponding coded V-BLAST system.

We also demonstrated the suitability of T-BLAST in a spatially correlated fading environment. In particular, we illustrated the importance of random interleavers for optimal performance of the RLST codes in a correlated channel environment. With sufficiently large random interleavers and receive antenna elements, the proposed RLST codes achieve near-optimal performance. Simulation results demonstrate the performance improvement that is possible by using larger interleaver sizes for T-BLAST architectures:

- We showed that when we have a large number of receive antennas, we can achieve near-optimal performance even in spatially correlated Rayleigh fading environments. The importance of independent and different time interleavers in achieving near-optimal performance and the existence of random interleavers in designing optimal RLST codes using 1-D linear channel codes were also demonstrated. A limitation of an ordinary BLAST architecture with interference cancellation receivers is that the performance degrades as the correlation between the transmit and receive antennas increases. By using simulation results, we showed that IDD receivers achieve near-optimal performance in correlated fading environments even though they use a form of parallel soft interference cancellation (PSIC) receivers. The reason for this improved performance is that the iterative decoder that uses the extrinsic and intrinsic information concepts inherent in the turbo principle is a close approximation of the global maximum likelihood decoding of the RLST codes.

- Developing a theoretical framework for the spectral efficiency of T-BLAST is a very difficult undertaking due to the nonlinear iterative receiver process after its second iteration. An empirical evaluation of the spectral efficiency achievable with specified channel codes and modulation has been provided. The evaluation was made using real-life data for both indoor and outdoor fixed wireless communication environments. In the indoor environment, due to the rich scattering process, we get an i.d.d. Rayleigh fading channel, but in the outdoor environments we have a correlated fading channel.

5.9 APPENDIX

Given (5.30) and (5.60), we wish to find the weight vectors $\mathbf{w}_k$ and u_k by minimizing the cost (convex) function

$$(\hat{\mathbf{w}}_k, \hat{u}_k) = \arg \min_{(\mathbf{w}_k, u_k)} E\{\|x_k - x_k^1\|^2\} \tag{5.76}$$

where the expectation is taken over noise and statistics of the data sequence.

Proof. The cost function is written as

$$C = E\{\|x_k^1 - x_k\|^2\} = E\{\|(\mathbf{w}_k^\dagger \mathbf{y} - u_k - x_k)\|^2\} \tag{5.77}$$

$$= \mathbf{w}_k^\dagger E\{\mathbf{y}\mathbf{y}^\dagger\}\mathbf{w}_k - \mathbf{w}_k^\dagger E\{\mathbf{y}(u+x)^*\} - E\{\mathbf{y}(u_k+x_k)^*\}^\dagger \mathbf{w}_k + E\{(u_k+x_k)^2\}$$

where

$$E\{\mathbf{y}\mathbf{y}^\dagger\} = E\left\{[\mathbf{h}_k x_k + \overline{\mathbf{H}}_k \mathbf{x}_k + \mathbf{v}][\mathbf{h}_k x_k + \overline{\mathbf{H}}_k \mathbf{x}_k + \mathbf{v}]^\dagger\right\}$$

$$= \mathbf{h}_k \mathbf{h}_k^\dagger + \overline{\mathbf{H}}_k E\left\{\mathbf{x}_k \mathbf{x}_k^\dagger\right\} \overline{\mathbf{H}}_k^\dagger + E\{\mathbf{v}\mathbf{v}^\dagger\} \tag{5.78}$$

and

$$E\{\mathbf{y}(u_k+x_k)^*\} = E\left\{[\mathbf{h}_k x_k + \overline{\mathbf{H}}_k \mathbf{x}_k + \mathbf{v}](u_k+x_k)^*\right\} = \mathbf{h}_k + \overline{\mathbf{H}}_k E\{\mathbf{x}_k\} u_k^* \tag{5.79}$$

Assuming that the soft outputs of different substreams are independent, we may write

$$E\left\{\mathbf{x}_k\mathbf{x}_k^\dagger\right\} = \mathbf{I}_{n_t-1} - \mathrm{Diag}[E\left\{\mathbf{x}_k\mathbf{x}_k^\dagger\right\}] + E\left\{\mathbf{x}_k\right\}E\left\{\mathbf{x}_k\right\}^\dagger \tag{5.80}$$

We use standard minimization techniques to solve the optimization problem formulated in (5.76). By setting $\frac{\partial C}{\partial u_k} = 0$ and $\frac{\partial C}{\partial \mathbf{w}_k} = 0$ and using (5.66), we get

$$u_k - \mathbf{w}_k^\dagger\overline{\mathbf{H}}_k E\left\{\mathbf{x}_k\right\} = 0$$
$$u_k = \mathbf{w}_k^\dagger\mathbf{z} \tag{5.81}$$

and

$$\left[\mathbf{h}_k\mathbf{h}_k^\dagger + \overline{\mathbf{H}}_k\left(E\{\mathbf{x}_k\mathbf{x}_k^\dagger\}\right)\overline{\mathbf{H}}_k^\dagger + E\left\{\mathbf{v}\mathbf{v}^\dagger\right\}\right]\mathbf{w}_k - \overline{\mathbf{H}}_k E\left\{\mathbf{x}_k\right\}u_k^* = \mathbf{h}_k$$
$$(\mathbf{P} + \mathbf{S} + \Sigma_{n_r})\mathbf{w}_k - \mathbf{z}u_k^* = \mathbf{h}_k \tag{5.82}$$

Solving (5.81) and (5.82), we finally get

$$\mathbf{w}_k = (\mathbf{P} + \mathbf{Q} + \Sigma_{n_r})^{-1}\mathbf{h}_k \tag{5.83}$$

which completes the proof of (5.62). $\qquad\qquad\qquad\qquad\qquad\qquad\qquad\square$

6

TURBO-MIMO SYSTEMS[1]

In Chapters 4 and 5, we studied different BLAST architectures and showed how the turbo principles applied to BLAST architectures can significantly improve overall system performance. In this chapter we discuss another important class of MIMO architecture, known as Turbo-MIMO systems, where the construction is based on space-time bit-interleaved coded modulation (ST-BICM). Turbo-MIMO is a highly effective system when used in conjunction with receivers employing iterative detection and decoding. This chapter also presents three recent low-complexity detection schemes that can be used in such systems: minimum-mean-squared-based soft interference cancellation (SIC-MMSE), iterative tree search (ITS) detection, and list-sphere detection. Simulation results are presented which show that, with the aid of these or similar schemes, the embodiment of the turbo processing principle in MIMO architectures provides a practical solution to the requirement of high data-rate transmission in a reliable manner for future wireless communications.

6.1 BIT-INTERLEAVED CODED MODULATION

The idea of bit-interleaved coded modulation (BICM), first introduced by Zehavi, can improve the performance of a coded modulation system over fading channels.

[1]This is an extended version of the paper by S. Haykin, M. Sellathurai, Y. L. C. de Jong, and T. Willink, "Turbo-MIMO for wireless communications," *IEEE Communications Magazine*, pp. 48–53, Oct. 2004. We acknowledge the important contributions of Drs. Yvo de Jong, Tharm Ratnarajah, and Tricia Willink to this chapter.

Space-Time Layered Information Processing for Wireless Communications,
By Mathini Sellathurai and Simon Haykin
Copyright © 2009 John Wiley & Sons, Inc.

The building blocks of the BICM encoding scheme are (1) a channel encoder, (2) an interleaver, and (3) a modulator. In particular, the performance of coded modulation over fading channels can be improved by introducing bitwise interleaving between the encoder and modulator at the transmit end. The receiving end will have (1) an appropriate soft-decision bit metric decoder, (2) an deinterleaver, and (3) the channel decoder matched to the encoding scheme used at the transmit end [21].

6.2 TURBO-MIMO THEORY AND ST-BICM

As already mentioned, Turbo-MIMO systems based on ST-BICM exploit the ideas of BICM and the turbo processing principle in a space-time coding framework. This BICM concept can be readily extended to MIMO channels, as shown in Figure 6.1, a block diagram of such a system with n_t transmit antennas and n_r receive antennas. In the following, it is assumed that $n_r \geq n_t$. Instead of employing dedicated space-time codes, as explained in Chapter 2, the transmitter uses a 1-D

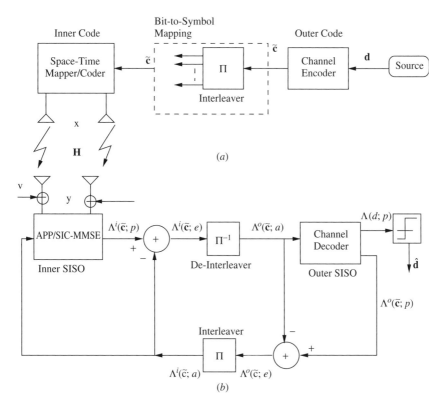

Figure 6.1 Block diagram of a MIMO system employing ST-BICM and an iterative receiver. Π and Π^{-1} denote interleaving and de-interleaving, respectively. The letters c and d refer to coded and uncoded bits, respectively.

forward error-correcting block code to encode the user's information bits. The channel encoder is followed by a pseudorandom interleaver, Π, and a space-time mapper, as discussed below. This configuration also can be viewed as a serial concatenation of two constituent encoders separated by the interleaver, the inner encoder being the space-time mapper working in conjunction with the outer channel encoder.

6.3 ST-BICM

Denoting a block of information bits by the vector $\mathbf{d}$ and the transfer function of the channel code by G, the codeword at the output of the outer encoder can be written as $\tilde{\mathbf{c}} = G\mathbf{d}$, and $\mathbf{c} = \Pi(\tilde{\mathbf{c}})$ represents the interleaved sequence of code bits, as shown in Figure 6.1. The modulation format is supposed to be identical for all transmit antennas. The number of bits per constellation point is denoted by M_c. The space-time mapper partitions $\mathbf{c}$ into L subvectors

$$\mathbf{c}^{(l)} = \left[c_{1,1}^{(l)}, \cdots, c_{1,M_c}^{(l)}, c_{2,1}^{(l)}, \cdots, c_{n_t,M_c}^{(l)} \right]^T, \quad l = 1, \cdots, L \tag{6.1}$$

and maps each of them onto a symbol vector

$$\mathbf{x}^{(l)} = \left[x_1^{(l)}, \cdots, x_{n_t}^{(l)} \right]^T \tag{6.2}$$

according to a unique predetermined bit-mapping scheme such that the symbol $x_j^{(l)}$ is mapped from the bit segment $c_{j,1}^{(l)}, \ldots c_{j,M_c}^{(l)}$. To simplify the notation, the superscript (l) will be omitted whenever no ambiguity arises. If $\mathbf{H} \in \mathcal{C}^{n_r \times n_t}$ is used to denote the MIMO channel, assumed to be narrowband, the received signal is given by

$$\mathbf{y} = \mathbf{Hx} + \mathbf{v} \tag{6.3}$$

where $\mathbf{v} \in \mathcal{C}^{n_r \times 1}$ is an additive noise vector whose elements are independent, complex-valued Gaussian variables with zero mean and variance σ^2.

In theory, it is possible to model the ST-BICM code, consisting of the channel encoder, space-time mapper, interleaver, and channel, as a single Markov process. However, the associated trellis representation is extremely complex and does not lend itself to computationally feasible decoding. The computational complexity of a full search procedure over such a trellis increases exponentially with the product of the number of transmit antennas n_t, the number of bits in a modulation M_c, and the code block length L. Fortunately, however, the interleaver separating the outer and inner encoders enables near-optimal decoding at reasonable computational complexity by exploiting the turbo processing principle. This approach, referred to as iterative detection and decoding (IDD), is similar to the decoding of serially concatenated turbo codes [139].

In general, the outer code of an ST-BICM MIMO system can use any type of error-correcting code that can be decoded with a SISO decoder, exemplified by convolutional or turbo codes. Turbo or turbo-like codes are often preferred because of their exceptionally good performance. The space-time mapper provides the flexibility to design the MIMO system so as to achieve the desired trade-off between multiplexing and diversity gain [26]. To maximumize the multiplexing gain, each antenna must transmit an independent information stream, i.e., the space-time mapper must be a spatial multiplexer. On the other hand, the space-time block codes discussed in Chapter 2 can be used to achieve the maximum diversity order.

6.4 ITERATIVE DETECTION AND DECODING

Iterative decoding of serially concatenated codes is now a well-established and practical alternative to optimal joint maximum-likelihood decoding following the material discussed in Chapter 4 on the turbo technique. Figure 6.1(b) shows the iterative receiver of an ST-BICM MIMO system, which separates the overall decoding problem into two stages: inner decoding (detection) and outer decoding (channel decoding). Information learned from one stage is passed on to the other stages iteratively; the process is continued until the receiver converges. The de-interleaver, Π^{-1}, is used to compensate for the interleaving operation used at the transmitter. Furthermore, working together with the interleaver, Π, the de-interleaver serves to decorrelate the output from one decoding stage before it is passed to the next. The iterative receiver updates and generally improves the soft decisions on the information bits as well as the code bits at each iteration of the message-passing process. These soft decisions are produced in the form of *a posteriori* LLRs, defined as

$$\Lambda(c) = \log \frac{P(c = +1|\cdot)}{P(c = -1|\cdot)} \tag{6.4}$$

In this equation, the probabilities $P(\cdot)$ are conditioned on the received signal vector **y** or the constraints of the channel code if the LLR is an output of the inner or outer decoder, respectively. Using Bayes' rule and assuming statistical independence between successive bits, which is a reasonable assumption because of the interleaving operation, the *a posteriori* LLR is reproduced here for convenience of presentation:

$$\underbrace{\log \frac{P(c = +1|\mathbf{y})}{P(c = -1|\mathbf{y})}}_{\Lambda(c;p)} = \underbrace{\log \frac{P(c = +1)}{P(c = -1)}}_{\Lambda(c;a)} + \underbrace{\log \frac{p(\mathbf{y}|c = +1)}{p(\mathbf{y}|c = -1)}}_{\Lambda(c;e)} \tag{6.5}$$

Here $\Lambda(c; a)$ constitutes the *a priori* information on bit c, and $\Lambda(c; e)$ constitutes the extrinsic information. This extrinsic information is the incremental new information learned from either the received signal vector or the channel code constraints, using the available *a priori* information. Similarly, the extrinsic information produced by the inner decoding stage is used as *a priori* information by the outer decoder, and vice versa.

With the above definitions, the iterative decoding algorithm, in its optimal implementation, has the following steps:

Step 1: The inner decoder, or detector, generates extrinsic LLRs for each of the $n_t M_c$ code bits mapped onto the symbol vector $\mathbf{x}$. This extrinsic information is written as

$$\Lambda^i(c_{jk}; e) = \log \left(\frac{\displaystyle\sum_{\mathbf{c} \in \mathbb{C}_{jk,+1}} \exp \mu(\mathbf{x})}{\displaystyle\sum_{\mathbf{c} \in \mathbb{C}_{jk,-1}} \exp \mu(\mathbf{x})} \right) - \Lambda^i(c_{jk}; a) \tag{6.6}$$

where $\mathbb{C}_{jk,\pm1}$ are the sets of all possible bit sequences of length $n_t M_c$ for which c_{jk} equals $+1$ and -1, respectively; that is,

$$\mathbb{C}_{jk,+1} = \{\mathbf{c} | c_{jk} = \pm1\} \tag{6.7}$$

The metric $\mu(\mathbf{x})$, included in (6.6) is given by

$$\mu(\mathbf{x}) = -\frac{1}{\sigma^2} \|\mathbf{y} - \mathbf{H}\mathbf{x}\|^2 + \sum_{i=1}^{n_t} \sum_{j \in \mathbb{J}_i} \Lambda^i(c_{ij}; a) \tag{6.8}$$

where

$$\mathbb{J}_i = \{j | j \in \{1, \cdots, M_c\} \text{ and } c_{ij} = +1\} \tag{6.9}$$

During the first iteration, all information bits are assumed equally likely to have been transmitted, i.e., $\Lambda^i(c_{ij}; a) = 0$ for all i, j. The detection algorithm based on (6.6)–(6.9) is known as the *a posteriori* probability (APP) or MAP algorithm.

Note that the inner decoder produces sets of $n_t M_c$ extrinsic LLR values for all L symbol vectors transmitted during a block period. The block of extrinsic LLR values associated with the bits in $\mathbf{c}$ is denoted by $\Lambda_{ext}^i(\mathbf{c})$; it becomes available to the outer decoder as *a priori* information after de-interleaving:

$$\Lambda^o(\tilde{\mathbf{c}}; a) = \Pi^{-1}\{\Lambda^i(\mathbf{c}; e)\} \tag{6.10}$$

Step 2: The outer decoder, in turn, processes the soft information $\Lambda^o(\tilde{\mathbf{c}}; a)$ and computes updated extrinsic information on both the information and code bits based on the trellis structure of the channel codes. The extrinsic information on the code bits, denoted by $\Lambda^o(\tilde{\mathbf{c}}; e)$, is reinterleaved and fed back to the inner decoder:

$$\Lambda^i(\mathbf{c}; a) = \Pi\{\Lambda^o(\tilde{\mathbf{c}}; e)\} \tag{6.11}$$

Steps 1 and 2 are repeated until convergence of the decoder is achieved. At that point, hard decisions on the information bits are made by taking the signs of the *a posteriori* LLR: $\Lambda^o(\mathbf{d}; p)$.

The complexity of the iterative receiver is determined by its two constituent decoders. If the channel code is a turbo-like code, the outer decoder is usually implemented as a concatenation of SISO modules based on the BCJR algorithm or approximations thereof. Refer to Chapter 4 for more details. As shown in (6.6), the inner detector performs an exhaustive search over the entire bit sequence of length $n_t M_c$, which can be mapped into a symbol vector $\mathbf{x}$. The associated complexity per information bit is therefore asymptotically proportional to $2^{n_t M_c}/n_t M_c$. Even for moderate numbers of transmit antennas and commonly used modulation formats, this complexity can be prohibitively high. In practice, therefore, suboptimal reduced-complexity MIMO detection schemes such as those discussed in the next section are often preferred.

6.5 SUBOPTIMAL MIMO DETECTION

6.5.1 List-Sphere Detection

Sphere detection (see [156]), also referred to as sphere decoding, is a reduced-complexity approximation to APP detection in which the search space is limited to the set of symbol vectors for which

$$\gamma(\mathbf{x}) = \|\mathbf{y} - \mathbf{Hx}\|^2 \le r^2$$

That is, the search is reduced to a set of symbols for which $\mathbf{Hx}$ lies within a hypersphere of some predetermined radius r centered on the noisy received signal ([75, 155]). The rationale for this approach is that symbol vectors for which $\|\mathbf{y} - \mathbf{Hx}\|^2$ is large are less likely to contribute significantly to the detector output (6.6) because their metrics, $\mu(\mathbf{x})$, defined in (6.8), are quite likely to be small enough to be negligible. These symbol vectors can therefore be excluded from the search space with limited degradation in performance. Because the reduced search space is, in general, only a small subset of the entire set of possibly transmitted signal points, the complexity of the sphere detector is considerably lower than that of the APP detector.

The procedure employed by the sphere detector to find the set of symbol vectors for which $\gamma(\mathbf{x}) \le r^2$ avoids an exhaustive search; rather, it is based on the fact that $\gamma(\mathbf{x})$ can be written as

$$\gamma(\mathbf{x}) = \|\mathbf{y} - \mathbf{Hx}\|^2 = (\mathbf{x} - \hat{\mathbf{x}})^{\dagger}\mathbf{H}^{\dagger}\mathbf{H}(\mathbf{x} - \hat{\mathbf{x}}) \tag{6.12}$$

where $\hat{\mathbf{x}} = [\hat{x}_1, \cdots, \hat{x}_{n_t}]^T = (\mathbf{H}^{\dagger}\mathbf{H})^{-1}\mathbf{H}^{\dagger}\mathbf{y}$ is the unconstrained maximum-likelihood solution. Since the matrix product $\mathbf{H}^{\dagger}\mathbf{H}$ is Hermitian and positive definite, it has a Cholesky decomposition $\mathbf{H}^{\dagger}\mathbf{H} = \mathbf{L}^{\dagger}\mathbf{L}$, in which $\mathbf{L} = [l_{ij}]$ is an $n_t \times n_t$ lower triangular matrix with real, positive diagonal entries. Using this Cholesky decomposition, (6.12) can be evaluated in a symbol-by-symbol manner. Starting with the first symbol x_1 and proceeding to x_{n_t}, we may write

$$\gamma_1 = |l_{11}(x_1 - \hat{x}_1)|^2$$

$$\gamma_i = \gamma_{i-1} + \left| l_{ii}(x_i - \hat{x}_i) + \sum_{j=1}^{i-1} l_{ij}(x_j - \hat{x}_j) \right|^2, \quad i = 2, \cdots, n_t \qquad (6.13)$$

$$\gamma(\mathbf{x}) = \gamma_{n_t}$$

and

$$\mu(\mathbf{x}) = -\frac{1}{\sigma^2}\gamma(\mathbf{x}) + \sum_{i=1}^{n_t}\sum_{j\in\mathbb{J}_i}\Lambda^i(c_{ij}; a) \qquad (6.14)$$

Because the terms γ_i are nonnegative, the search criterion $\gamma(\mathbf{x}) \le r^2$ implies that $\gamma_i \le r^2$ for $i = 1, \cdots, n_t$. Hence, the sphere detection algorithm starts by selecting the set of symbols x_1 for which $\gamma_1 \le r^2$. For each of these symbols, the algorithm subsequently selects the symbols x_2 for which $\gamma_2 \le r^2$, and this procedure continues until the condition $i = n_t$ is reached. The symbol-by-symbol estimation algorithm expressed by (6.49) is derived for three transmit antennas as an example in Appendix 6.1.

In [73], Hotchwald and ten Brink propose a list version of this scheme, called the list sphere detector (LSD). In this scheme, only the M symbol vectors that lie closest to the received signal vector, collectively referred to as the candidate list, $\mathbb{L}$, are used to produce the detector output. Mathematically, the detector output is described by

$$\Lambda^i(c_{jk}; e) = \log\left(\frac{\displaystyle\sum_{\mathbf{c}\in(\mathbb{L}\cap\mathbb{C}_{jk,+1})} \exp\mu(\mathbf{x})}{\displaystyle\sum_{\mathbf{c}\in(\mathbb{L}\cap\mathbb{C}_{jk,-1})} \exp\mu(\mathbf{x})}\right) - \Lambda^i(c_{jk}; a) \qquad (6.15)$$

Note that, according to Steingrimsson, Luo, and Wong [144], it is reported that the complexity per bit of LSD is still exponential in $N_t M_c$. For $M = 2^{n_t M_c}$ and sufficiently large r, the LSD scheme is identical to that of the APP detector. The complexity per bit of this scheme is roughly quadratic in n_t, and its dependence on the constellation size is asymptotically proportional to $2^{M_c}/M_c$. It is noted that, in practice, the log-sum over exponential functions, which is a relatively complex operation, may be approximated by

$$\log\sum_j \exp\mu_j \approx \max_j \mu_j \qquad (6.16)$$

with little performance degradation. This approximation was discussed in Chapter 4. At each iteration, the quality of the candidate list may be improved by taking into account the *a priori* information computed in the previous iteration. This is utilized in the ITS detection scheme, where an improved candidate list is produced at each iteration based on the knowledge of LLRs, as proposed in [35]. It has been shown that a considerably better performance than that of the LSD scheme can be achieved at the same list sizes.

6.5.2 ITS Detection

The ITS detection scheme proposed in [35] has better performance than the LSD, as it produces an improved candidate list at each iteration of the receiver by taking into account the *a priori* information fed back from the outer decoder. This improved candidate list is generated with the aid of a breadth-first tree search algorithm known as the M-algorithm, as illustrated in Figure 6.2. The set of all possible symbol vectors can be represented by a tree structure of depth n_t, having a single symbol on each branch and 2^{M_c} branches emanating from each node. Each path within the tree is uniquely associated with a sequence of symbols $s_1, \cdots, s_d$ and a metric μ_d, where $d \le n_t$ indicates the depth of the path and μ_d is defined by

$$\mu_1 = -\frac{1}{\sigma^2}|l_{11}(x_1 - \hat{x}_1)|^2 + \sum_{j \in \mathbb{J}_1} \Lambda(c_{1j}; a)$$

$$\tag{6.17}$$

$$\mu_i = \gamma_{i-1} + -\frac{1}{\sigma^2}\left|l_{ii}(x_i - \hat{x}_i) + \sum_{j=1}^{i-1} l_{ij}(x_j - \hat{x}_j)\right|^2 + \sum_{j \in \mathbb{J}_i} \Lambda(c_{ij}; a), \quad i = 2, \cdots, n_t$$

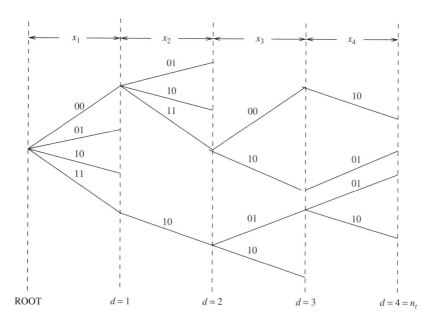

Figure 6.2 Example of a sequential tree search for $n_t = 4$, $M_c = 2$. At each symbol depth, the best $M = 4$ paths are retained. Deleted paths are not shown.

Every possible symbol vector corresponds to a path to the maximum depth and has an associated metric $\mu(\mathbf{x}) = \mu_{n_t}$, as defined in (6.8).

The ITS detector uses the M-algorithm to search for the M best paths through the tree. The M-algorithm keeps only the M paths in the tree with the largest metrics. At each depth smaller than n_t, the algorithm keeps a candidate list, $\mathbb{L}$, of the best M paths found thus far and moves forward by extending each of these paths to form $M \cdot 2^{M_c}$ new paths where M_c is the constellation size. Path metrics are then updated with the aid of (6.17), and only the M best new paths are retained in the updated list; the remaining $M(2^{M_c} - 1)$ paths are deleted. When the algorithm reaches the maximum depth, the candidate list, $\mathbb{L}$, is used to compute an approximation of the APP detector output, using (6.15) or the max-log approximation.

Theoretically, the performance of the ITS detector is identical to that of the APP detector only for the maximum possible list size $2^{n_t M_c}$. In practice, however, near-optimum performance is often achieved when M is only a small fraction of the full search space. The complexity per bit of the ITS detector is hardly dependent on n_t and, as with the LSD, its dependence on the signal constellation size is of order $2^{M_c}/M_c$.

6.5.3 Multilevel Mapping ITS Detection

The exponential growth in complexity of the ITS detector with increasing M_c is a consequence of the fact that the number of branches from each node in the tree structure is 2^{M_c}. Consequently, the number of metric updates at each step of the tree search is proportional to 2^{M_c}. This rapid complexity increase is a disadvantage if higher order modulation schemes are employed, which are of special interest if high spectral efficiency is desired. Fortunately, the use of multilevel QAM signal constellations can significantly reduce the complexity of the tree search [35].

A QAM constellation with multilevel bit mapping can be partitioned into four equal subsets such that (1) the maximum Euclidean distance between the signal points in each subset is minimized, (2) each subset is uniquely identified by the first 2 bits of its signal points, and (3) the remaining $M_c - 2$ bits of each subset again form a new multilevel mapping. An example of a 64-QAM signal constellation with multilevel Gray bit mapping is shown in Figure 6.3. As indicated by the dotted lines, this constellation can be successively partitioned into square subsets with a minimum mean intrasubset Euclidean distance. At each partitioning level l, the $2l$ most significant bits of a signal point determine in which subset it is located. Because the subsets form dense clusters of signal points, decisions on the first bit pairs in the label can be made without considering the subsequent bits; this is done by selecting the subset whose center of gravity best matches the received signal.

In the multilevel map-ITS detection scheme, the breadth-first tree search is performed in steps of 2 bits at a time, even if $M_c > 2$. At any stage, the surviving paths are extended according to the four possible values of the next, say lth, bit pair in the label of the current symbol x_i. The "intermediate" metric associated with each of these extended paths is evaluated from (6.17), in which x_i is replaced by the center of gravity of the corresponding constellation subset at the lth partitioning

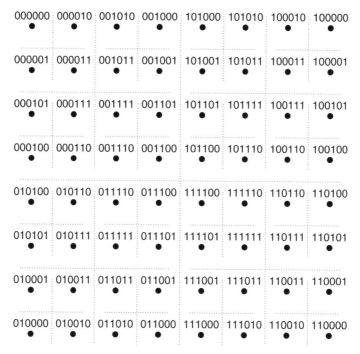

Figure 6.3 Example of a 64-QAM signal constellations with multilevel Gray bit mapping.

level. Furthermore, the *a priori* LLRs $\Lambda^i(c_{ij}, a)$ with $j > 2l$ are replaced by zeros in the evaluation of (6.17). Similar to the ITS detector, after each step, only the M largest intermediate metrics are retained. When the paths departing from the previous symbol depth have been extended by M_c bits, they reach the next symbol depth. At this point, the actual 2^{M_c}-ary QAM signal points can be used in the computation of the path metrics.

Because the number of metric updates per bit in the multilevel map-ITS detection scheme is $2M$, instead of $M \cdot 2^{M_c}/M_c$ in the basic ITS scheme, its complexity per bit is practically independent of the alphabet size. Moreover, computer simulations have shown that the complexity reduction due to the use of multilevel mapping does not lead to any performance degradation relative to the basic scheme.

6.5.4 Soft Interference Cancellation MMSE Detection

The soft interference cancellation minimum mean squared error (SIC-MMSE) detector proposed in [132] treats each of the symbols in the symbol vector $\mathbf{x}$ as being distorted by the other $n_t - 1$ symbols in $\mathbf{x}$ and noise. It uses the *a priori* information available at its input to estimate and cancel the interference due to the other symbols, then suppresses the residual interference and noise with the aid of the MMSE criterion. This scheme was originally developed for constant-envelope modulation, but it can be also be extended to the QAM modulation scheme [33].

Let x_j be the desired symbol; then (6.3) can be rewritten as

$$\mathbf{y} = \underbrace{\mathbf{h}_j x_j}_{\text{desired signal}} + \underbrace{\sum_{i \neq j} \mathbf{h}_i x_i}_{\text{interference}} + \underbrace{\mathbf{v}}_{\text{noise}} \tag{6.18}$$

where $\mathbf{h}_j \in C^{n_r \times 1}$ denotes the jth column vector of channel matrix $\mathbf{H}$. The SIC-MMSE scheme produces an estimate of x_j by subtracting the estimated interference due to the other symbols from $\mathbf{y}$ and suppressing the residual interference and noise by means of adaptive linear filtering as follows:

$$\hat{x}_j = \mathbf{w}_j^{\dagger} \left(\mathbf{y} - \sum_{i \neq j} \mathbf{h}_i \hat{x}_i \right) \tag{6.19}$$

Estimates of the interfering symbols involved in the summation term of (6.19) are obtained as

$$\hat{x}_i = \mathcal{E}\{x_i\} \tag{6.20}$$

The corresponding mean-square estimation errors are denoted by

$$\sigma_{\hat{x},i}^2 = \mathcal{E}\{|\hat{x}_i - x_i|^2\} \tag{6.21}$$

In (6.20) and (6.21), the expectation $\mathcal{E}\{\cdot\}$ is taken over the 2^{M_c} possible realizations of x_i, weighted by their respective *a priori* probabilities, which are determined from $\Lambda^i(\mathbf{c}; a)$. The optimal weighting vector $\mathbf{w}_{j,opt} \in C^{n_r \times 1}$ minimizes the mean-square error in the estimate of x_j after soft interference cancellation and linear filtering; that is,

$$\mathbf{w}_{j,opt} = \arg \min_{\mathbf{w}_j} E\{|x_j - \hat{x}_j|^2\} \tag{6.22}$$

The solution to this optimization problem can be obtained by standard minimization techniques; it is given by

$$\mathbf{w}_{j,opt} = \left(\mathbf{h}_j \mathbf{h}_j^{\dagger} + \sum_{i \neq j} (\sigma_{\hat{x},i}^2 / \sigma_x^2) \mathbf{h}_i \mathbf{h}_i^{\dagger} + (\sigma^2 / \sigma_x^2) \mathbf{I}_{n_r} \right)^{-1} \mathbf{h}_j \tag{6.23}$$

where σ_x^2 is the average power used on each of the transmit antennas. In the derivation of (6.23), we utilize the fact that all symbols are statistically independent of the noise and of each other. Proof of (6.23) is presented in Appendix 6.2.

Based on the assumption that the error $\hat{x}_j - x_j$ is a zero-mean complex Gaussian variable and is statistically independent of the errors on the other symbols in $\mathbf{x}$, the SIC-MMSE detector approximates the APP detection output (6.6) as

$$\Lambda(c_{jk}; e) = \log \left(\frac{\sum\limits_{\mathbf{c}_j \in \mathbb{C}_{jk,+1}} \exp \zeta(x_j)}{\sum\limits_{\mathbf{c}_j \in \mathbb{C}_{jk,-1}} \exp \zeta(x_j)} \right) - \Lambda(c_{jk}; a) \tag{6.24}$$

in which $\mathbf{c}_j = [c_{j1}, \cdots, c_{jM_c}]^T$ is the vector of bits mapped onto x_j, and

$$\zeta(x_j) = -\frac{|\hat{x}_j - x_j|^2}{\sigma_x^2(1 - \mathbf{h}_j^\dagger \mathbf{w}_j)} + \sum_{l \in \mathbb{J}_j} \Lambda^i(c_{jl}; a) \qquad (6.25)$$

In (6.24), $\mathbb{C}_{jk,\pm 1}$ has a slightly different definition from that in (6.7), as shown by

$$\mathbb{C}_{jk,\pm 1} = \{\mathbf{c}_j | c_{jk} = \pm 1\} \qquad (6.26)$$

The log-sum function in (6.24) can be approximated using the max-log approximation discussed in Chapter 4.

For large n_r, the complexity of the SIC-MMSE detector is dominated by the inversion of the Hermitian positive definite matrix in (6.23) and is on the order of n_r^3. For large M_c, on the other hand, the overall complexity is dominated by the evaluation of (6.20) and (6.21), and the complexity per bit is on the order of $2_c^M / M_c$.

6.6 SIMULATION FOR NARROWBAND TURBO-MIMO

In this section, results are provided for two sets of computer simulations of an ST-BICM MIMO system employing an iterative receiver. The first simulation illustrates the BER performance gain achieved by using turbo codes compared to convolutional codes as outer codes. The second simulation provides a comparison of BER performance for the suboptimal reduced-complexity inner decoding schemes described in the previous section.

Experiment 1: Convolutional versus Turbo Codes The outer codes considered in the first simulation are 8-state, rate-1/2 turbo and convolutional codes. The turbo codes are composed of two identical 8-state convolutional codes generated by using a recursive FB generator polynomial [1011] and a FF polynomial [1101]. No effort was made to optimize either the pseudorandom interleaver used in the turbo encoder and decoder or the interleaver used to separate the channel encoder from the space-time mapper. The antenna configuration considered here is $n_t = n_r = 4$. The modulation format is QPSK, and the MIMO detection scheme is SIC-MMSE. Both real-life measurements, taken in Manhattan using a (16,16) MIMO testbed [56], and a theoretical model were used to generate the MIMO channel coefficients. The theoretical model generates temporally and spatially uncorrelated complex Gaussian distributed channel coefficients at each symbol interval. In the following two examples, the transmission is organized in blocks of 3200 information bits. The maximum number of iterations of the turbo decoder is restricted to 10 iterations. The ratio E_b/N_0 denotes the average SNR at each receive antenna.

Figure 6.4 shows BER versus SNR based on the assumption that a turbo-encoded block is formed by independent channel samples at each symbol interval. The results were obtained by using real-life (dashed lines) and ideal MIMO channel

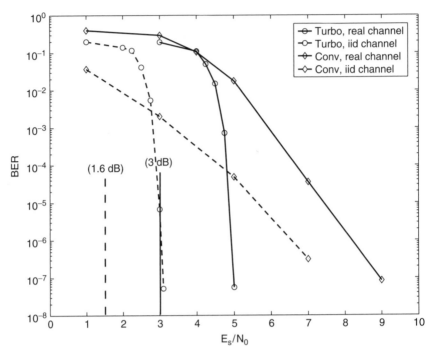

Figure 6.4 Error performance of a 4 × 4 ST-BICM MIMO system employing the SIC-MMSE detectors, using 8-state turbo and convolutional codes as channel codes. The modulation format is QPSK.

data (solid lines). In this simulation, we used 5 outer (detector-decoder) iterations with inner (turbo decoding) iterations of 1, 2, 4, 8, and 10 per outer iteration, respectively. For comparison, we also show the corresponding results with 8-state convolutional codes used as outer codes. The figure shows that more than 4 dB gain is achieved at BER $= 10^{-6}$ over the ST-BICM system with convolutional codes. Note that the performance with the ideal MIMO channel is 1–2 dB superior to the one with the real channel. This is because the capacity of a real channel is lower than that of an ideal MIMO channel due to possible spatial and temporal correlations. In the figure, ergodic MIMO capacity limits [45] of the 4 × 4 MIMO channels (for both independent Gaussian channel coefficients and real-life measured channels) are indicated by vertical lines. Perfect knowledge of channel state information (CSI) is assumed in the capacity calculations.

Note: Given the same code rate and the number of states, a turbo code has much higher decoding complexity than a convolutional code. It has been found in [23] that under this condition, turbo codes show performance advantages over convolutional codes only in the AWGN channel or when the order of diversity (either in space, time, or frequency) is sufficiently high. Reference [23] is a good reference on this topic. The complexity of various decoding schemes for turbo codes and convolutional codes is quantified in Tables I–IV of [23].

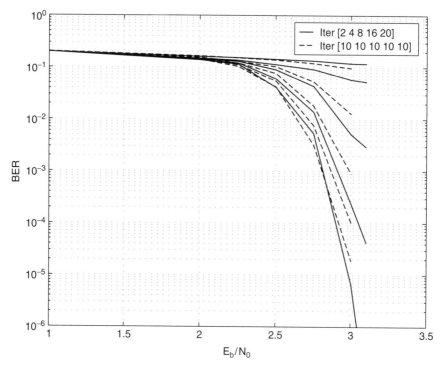

Figure 6.5 Error performance of a 4 × 4 ST-BICM SIC-MMSE system. QPSK modulation. The channel code is an 8-state turbo code of average rate 1/2.

Experiment 2: Performance versus Detector-Decoder Iterations with 8-State Turbo Codes as Outer Channel Codes Figure 6.5 shows the BER versus SNR for one, two, three, four, and five detector-decoder iterations of the SIC-MMSE receiver for two different sets of decoder (turbo decoding) iterations. This simulation assumes an i.i.d. Gaussian channel. For both sets, receiver performance improves significantly with increasing number of detector-decoder iterations. When we use the set of decoder (turbo-code) iterations (2, 4, 8, 16, and 20 per outer iteration), the receiver tends to perform better compared to the other set (10, 10, 10, 10, and 10 per outer iteration). However, after a few iterations, due to the feedback of correlated noises, the iterative gain is reduced.

Experiment 3: Comparison Between Suboptimal Detection Schemes The following computer simulations provide a comparison of BER performance for the reduced-complexity MIMO detection schemes discussed in the previous section. Again, the antenna configuration is $n_t = n_r = 4$. The channel code is a turbo code of rate-1/2 and memory 2, whose FF and FB generators are [101] and [111], respectively. The elements of the channel matrix are samples of independent, complex-valued, zero-mean Gaussian processes. The channel is

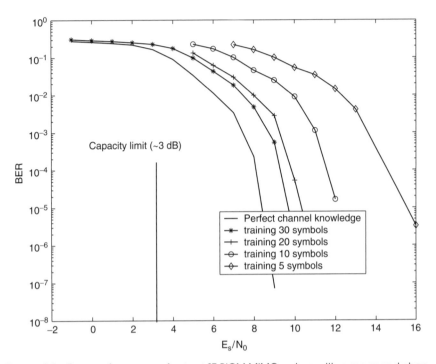

Figure 6.6 Error performance of a 4 × 4 ST-BICM MIMO system with a measured channel under block fading conditions; estimated CSI with different lengths of training symbols. An 8-state turbo code is used as the outer channel code with QPSK modulation.

assumed to remain constant over blocks of 192 information bits, and every new channel realization is statistically independent of previous blocks. The block length of the channel code, and therefore the size of the interleaver Π, is equal to 18432 code bits. The number of iterations in the detector-decoder loop is four, and that of the turbo decoder is limited to eight per outer iteration. All interleavers are pseudorandom; no attempt was made to optimize their design. The modulation formats considered are QPSK, 16-QAM and 64-QAM, all with multilevel Gray mapping.

In Figure 6.6, the performance of a 4 × 4 ST-BICM MIMO system with perfect knowledge of CSI is compared to that of the system with estimated CSI using a pilot-aided channel estimation algorithm. As expected, system performance improves as the number of training symbols increases. With 30 training symbols, the performance loss incurred by the channel estimation error is about 1 dB at the target BER = $10^{-3}/10^{-4}$ compared to the genie-aided case with perfect CSI. The performance gap can be further reduced by transmitting pilots at the expense of reduced spectral efficiency.

Figure 6.7 shows BER performance versus E_s/N_0 for LSD, multilevel map-ITS, and SIC-MMSE detection. The performance of the basic ITS detection scheme is identical to that of multilevel map-ITS and is not shown in the figure. Ergodic

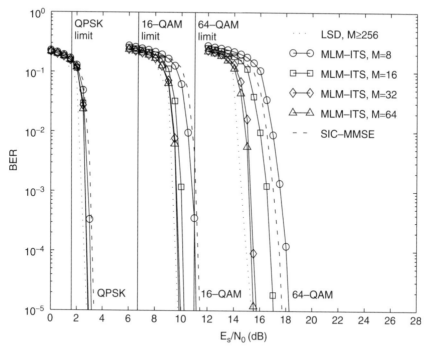

Figure 6.7 Error performance of a 4 × 4 ST-BICM MIMO system employing the LSD, the multilevel map-ITS, and the SIC-MMSE detector. The channel code is a 4-state turbo code of rate 1/2.

MIMO capacity bounds are shown for reference. On the basis of the results presented in the figure, we may make the following observations:

- As expected, the performance of multilevel map-ITS improves if the list size M is increased, in this case from 8 to 64.
- The performance of multilevel map-ITS can be made comparable to that of SIC-MMSE by tuning the list size. Interestingly, as shown in [34], the complexities of both schemes are comparable at the same performance level. However, an advantage of multilevel map-ITS is that its performance can be traded off for lower complexity.
- To achieve comparable error performance, the list size of the LSD scheme must be chosen to be approximately eight times larger than that of multilevel map-ITS. Note that the list size of the LSD scheme is 256 for QPSK and 512 for 16-QAM and 64-QAM.
- The ST-BICM system with the best performance shown in Figure 6.7 (LSD) operates at approximately 1, 4, and 6 dB from the capacity limit (at a BER of 10^{-5}) for QPSK, 16-QAM, and 64-QAM, respectively.

6.7 WIDEBAND TURBO-MIMO (ST-BICM)

For high-date-rate applications, such as high-speed downlink packet access (HSDPA), it may be necessary to utilize signals whose bandwidth exceeds the coherence bandwidth of the wireless channel, which raises the issue of frequency-selective channels. The problem can be tackled by employing orthogonal frequency division multiplexing (OFDM) technology, which transforms the frequency-selective channel into a number of parallel flat fading channels. Another effective remedy to combat the detrimental effects of the frequency-selective channel is the use of equalization. We focus on the latter approach in this section and extend the Turbo-MIMO concepts to wideband Turbo-MIMO systems.

The ST-BICM scheme (Figure 6.1) and its corresponding iterative detection and decoding receiver (Figure 6.2) can also be exploited to design efficient, low-complexity tranceivers for MIMO frequency-selective wireless channels. However, the front-end component of the iterative reciever would have to be designed so as to equalize the intersymbol interferences (ISI) produced by the channel. In particular, the receiver structure is made up of the following constituents:

- A SISO channel equalizer designed to mitigate the effect of ISI produced by the transmission of the encoded-interleaved signal across the frequency-selective channels; the equalizer acts as the inner decoder.
- A SISO channel decoder designed to improve the estimates of encoded data symbols; the channel decoder acts as the outer decoder.

The optimal inner decoder is a MAP decoder based on the decoding trellis, formed on the basis of the channel impulse response (i.e., the tap weights of the tapped-delay-line model). However, due to the exponential complexity of the optimal inner decoders, sub-optimal inner decoders are often preferred. We will describe two such equalizers based on MMSE estimation and iterative trellis search algorithms.

The frequency-selective channel can be modeled as an FIR (i.e., tapped-delay-line) filter whose impulse response length is denoted by N_h. The sampled channel response from transmitter i to receiver j is denoted by the vector $\mathbf{h}_{ij} = \left[h_{ij}(0), h_{ij}(1), \ldots, h_{ij}(N_h - 1) \right]^T$, which includes the effect of transmit and receive filters and no time dependence within every individual burst. If we denote the discrete-time index by k, then the signal vector received at the channel output can be written as

$$\mathbf{y}(k) = \sum_{l=0}^{N_h - 1} \mathbf{H}(l)\mathbf{x}(k - l) + \mathbf{v}(k) \in \mathbb{C}^{n_r \times 1} \qquad (6.27)$$

where $\mathbf{x}(k) = [x_1(k), \ldots, x_{n_t}(k)]^T$ is the $n_t \times 1$ transmitted signal vector and $\mathbf{v}(k) = [v_1(k), \ldots, v_{n_r}(k)]^T$ is the $n_r \times 1$ channel noise modeled as zero-mean

AWGN, i.e., $v \sim \mathcal{CN}(0, \sigma^2 I_{n_r})$, and

$$
\mathbf{H}(l) = \begin{bmatrix} h_{11}(l) & \cdots & h_{n_t 1}(l) \\ \vdots & \vdots & \vdots \\ h_{1n_r}(l) & \cdots & h_{n_t n_r}(l) \end{bmatrix}, \quad l = 0, \ldots, N_h - 1
$$

is the $n_r \times n_t$ time-varying channel matrix channel impulse response.

Equivalently, we may express the received signal sample at antenna j and discrete-time k as

$$
y_j(k) = \sum_{i=1}^{n_t} \sum_{l=0}^{N_h-1} h_{ij}(l) x_i(k - l) + v_j(k) \tag{6.28}
$$

where $h_{ij}(l)$ is the channel response of the ith transmitter to the jth receiver path at time $l = 0, 1, \ldots, N_h - 1$, $x_i(k)$ is the ith antenna's transmitted symbol at time k, and $v_j(k)$ is the corresponding additive channel noise.

The signal model of (6.28) can be extended to a stacked block data model by stacking $N_s + N_h - 1$ received signals of $y_j(k)$'s into an $(N_s + N_h - 1) \times 1$ vector $\mathbf{y}_j = [y_j(1), \ldots, y_j(N_s + N_h - 1)]^T$. This stacked received signal at antenna j may thus be written as

$$
\mathbf{y}_j = \mathbf{H}_j \mathbf{x} + \mathbf{v}_j \in \mathbb{C}^{(N_s + N_h - 1) \times 1} \tag{6.29}
$$

where $\mathbf{x} = \text{vec}([\mathbf{x}_1 \cdots \mathbf{x}_{n_t}])$, $\mathbf{x}_i = [x_i(1), \ldots, x_i(N_s)]^T$, $\mathbf{v}_j = [v_j(1), \ldots, v_j(N_s + N_h - 1)]^T$. The matrix $\mathbf{H}_j = [\mathbf{H}_{1j}, \mathbf{H}_{2j}, \ldots, \mathbf{H}_{n_t j}] \in \mathbb{C}^{(N_s + N_h - 1) \times N_s n_t}$ has a block-Toeplitz structure with

$$
\mathbf{H}_{ij} = \text{diag}(\mathbf{h_{ij}}, \ldots, \mathbf{h_{ij}}) \in \mathbb{C}^{(N_s + N_h - 1) \times N_s}
$$

Correspondingly, the n_r antenna received signals are stacked as shown by

$$
\mathbf{y} = \sum_{j=1}^{n_r} [\mathbf{H}_j]^\dagger \mathbf{y}_j = \sum_{j=1}^{n_r} [\mathbf{H}_j]^\dagger [\mathbf{H}_j] \mathbf{x} + [\mathbf{H}_j]^\dagger \mathbf{v}_j
$$
$$
= \mathbf{R}\mathbf{x} + \mathbf{v} \in \mathbb{C}^{N_s n_t \times 1} \tag{6.30}
$$

where

$$
\mathbf{R} = \sum_{j=1}^{n_r} [\mathbf{H}_j]^\dagger [\mathbf{H}_j] \tag{6.31}
$$

and

$$
\mathbf{v} = \sum_{j=1}^{n_r} [\mathbf{H}_j]^\dagger \mathbf{v}_j \tag{6.32}
$$

The covariance matrix of the additive noise vector $\mathbf{v}$ is given by

$$\mathbf{R}_{vv} = \mathcal{E}[\mathbf{v}\mathbf{v}^\dagger] = \mathcal{E}_v \left\{ \left(\sum_{j=1}^{n_r} [\mathbf{H}_j]^\dagger \mathbf{v}_j \right) \left(\sum_{j=1}^{n_r} [\mathbf{H}_j]^\dagger \mathbf{v}_j \right)^\dagger \right\}$$

$$= \sigma^2 \sum_{j=1}^{n_r} [\mathbf{H}_j]^\dagger [\mathbf{H}_j] = \sigma^2 \mathbf{R} \tag{6.33}$$

where the expectation is taken with respect to the noise statistics. In deriving (6.33), we have made two assumptions:

- The channel matrices $\mathbf{H}_j$ are constant matrices.
- The elememnts of noise vector $\mathbf{v}$ are statistically independent, that is,

$$\mathcal{E}_v \left\{ [\mathbf{v}_j][\mathbf{v}_i]^\dagger \right\} = \begin{cases} \sigma^2 \mathbf{I} & i = j \\ 0 & i \neq j \end{cases}$$

6.7.1 MIMO Equalizer

Based on the model expressed by (6.30), we can derive equalizers, depending on the criterion of interest. Under the zero-forcing (ZF) criterion, the soft estimates of transmitted signals are given by

$$\hat{\mathbf{x}}_{ZF} = \arg\min_{\mathbf{x}} \left\{ (\mathbf{y} - \mathbf{R}\mathbf{x})^\dagger \mathbf{R}_{vv}^{-1} (\mathbf{y} - \mathbf{R}\mathbf{x}) \right\}$$

$$= \left(\mathbf{R}^\dagger \mathbf{R}_{vv}^{-1} \mathbf{R} \right)^{-1} \mathbf{R}^\dagger \mathbf{R}_{vv}^{-1} \mathbf{y} \tag{6.34}$$

$$= \mathbf{R}^{-1} \mathbf{y} \in \mathbb{C}^{N_s n_t \times 1}$$

On the other hand, under the MMSE criterion, the soft estimates of the transmitted signals are given by

$$\hat{\mathbf{x}}_{MMSE} = \arg\min_{\mathbf{x}} \mathcal{E} \left\{ (\hat{\mathbf{x}} - \mathbf{x})^\dagger (\hat{\mathbf{x}} - \mathbf{x}) \right\}$$

$$= \left(\mathbf{R}^\dagger \mathbf{R}_{vv}^{-1} \mathbf{R} + \mathbf{R}_{xx}^{-1} \right)^{-1} \mathbf{R}^\dagger \mathbf{R}_{vv}^{-1} \mathbf{y} \tag{6.35}$$

$$= \left(\mathbf{R} + \sigma^2 \mathbf{I} \right)^{-1} \mathbf{y} \in \mathbb{C}^{N_s n_t \times 1}$$

For an ill-conditioned $\mathbf{R}$ matrix and at low SNRs, the MMSE equalizer performs significantly better than the ZF equalizer. Next, we consider the details of an MMSE-based joint equalization and soft interference cancellation receiver.

MMSE-Based Joint Equalization and Soft Interference Cancellation

Here we describe the SIC-MMSE technique. Let $\mathbf{x}_i \in \mathbb{C}^{N_s \times 1}$ be the desired signal. The channel matrix for the interference signal $\mathbf{x}_{int} = \text{vec}([\mathbf{x}_1 \cdots \mathbf{x}_{i-1}\mathbf{x}_{i+1} \cdots \mathbf{x}_{n_t}])$ is given by $\mathbf{H}_{int,j} = [\mathbf{H}_{1j}, \dots, \mathbf{H}_{i-1j}, \mathbf{H}_{i+1j}, \dots, \mathbf{H}_{n_t j}]$. The stacked received signals at antenna j due to the data block of N_s transmitted data symbols can be written as

$$\mathbf{y}_j = \mathbf{H}_{ij}\mathbf{x}_i + \mathbf{H}_{int,j}\mathbf{x}_{int} + \mathbf{v}_j \in \mathbb{C}^{(N_s+N_h-1)\times 1} \qquad (6.36)$$

The stacked version of n_r antenna received signals is

$$\mathbf{y} = \sum_{j=1}^{n_r}[\mathbf{H}_{ij}]^{\dagger}\mathbf{y}_j = \mathbf{A}_i\mathbf{x}_i + \mathbf{B}_i\mathbf{x}_{int} + \mathbf{u}_i \in \mathbb{C}^{N_s \times 1} \qquad (6.37)$$

where

$$\mathbf{A}_i = \sum_{j=1}^{n_r}[\mathbf{H}_{ij}]^{\dagger}[\mathbf{H}_{ij}] \in \mathbb{C}^{N_s \times N_s},$$

$$\mathbf{B}_i = \sum_{j=1}^{n_r}[\mathbf{H}_{ij}]^{\dagger}\mathbf{H}_{int,j} \in \mathbb{C}^{N_s \times N_s(n_t-1)}$$

and

$$\mathbf{u}_i = \sum_{j=1}^{n_r}[\mathbf{H}_{ij}]^{\dagger}\mathbf{v}_j$$

Let $\mathbf{W}_i$ be an $N_s \times N_s$ weight matrix used to estimate the desired signal $\mathbf{x}_i$ from (6.37). The SIC-MMSE scheme produces an estimate of $\mathbf{x}_i$ by subtracting the estimated interference due to the other symbols from $\mathbf{y}$ and suppressing the residual interference and noise by means of adaptive linear filtering, i.e.,

$$\hat{\mathbf{x}}_i = \mathbf{W}_i^{\dagger}(\mathbf{y} - \mathbf{B}_i\mathbf{x}_{int}) \qquad (6.38)$$

Estimates of the interfering symbols are obtained as

$$\hat{x}_j = \mathcal{E}\{x_j\} \qquad (6.39)$$

and the corresponding mean-squared estimation errors are denoted by

$$\sigma_{\hat{x},j}^2 = \mathcal{E}\{|x_j - \hat{x}_j|^2\} \qquad (6.40)$$

In (6.39) and (6.40), the expectation $\mathcal{E}\{\cdot\}$ is taken over the 2^{M_c} possible realizations of x_i weighted by their respective *a priori* probabilities, which are determined from $\Lambda^i(\mathbf{c}; a)$. The optimal weighting vector $\mathbf{W}_{i,opt} \in \mathcal{C}^{n_r \times 1}$ minimizes the mean-squared error in the estimate of $\mathbf{x}_i$ after soft interference cancellation and linear filtering, i.e.,

$$\mathbf{W}_{i,opt} = \arg\min_{\mathbf{W}_i} E\{|\mathbf{x}_i - \hat{\mathbf{x}}_i|^2\} \qquad (6.41)$$

Using standard optimization techniques, the MMSE solution to (6.37) is given by

$$\hat{\mathbf{W}}_i = \left(\mathbf{A}_i \mathbf{A}_i^\dagger + \frac{\mathbf{B}_i \mathcal{E}\{(\mathbf{x}_{int} - \hat{\mathbf{x}}_{int})(\mathbf{x}_{int} - \hat{\mathbf{x}}_{int})\}\mathbf{B}_i^\dagger}{\sigma_x^2} + \frac{\sigma^2}{\sigma_x^2}\mathbf{A}_i \right)^{-1} \mathbf{A}_i$$

With an increasing number of iterations, we assume that in the limit, $\hat{\mathbf{x}}_{int}$ converges to the true value $\mathbf{x}_{int}$, in which case we may ignore the term $\mathcal{E}\{(\mathbf{x}_{int} - \hat{\mathbf{x}}_{int})(\mathbf{x}_{int} - \hat{\mathbf{x}}_{int})^\dagger\}$. Then the MMSE soft estimation of the desired signal is given by

$$\hat{\mathbf{x}}_{i,\text{MMSE}} = \left(\sum_{j=1}^{n_r} [\mathbf{H}_{ij}]^\dagger [\mathbf{H}_{ij}] + (\sigma^2/\sigma_x^2)\mathbf{I} \right)^{-1} \mathbf{y} \in \mathbb{C}^{N_s \times 1} i = 1, \ldots, n_t. \quad (6.42)$$

See Appendix 6.2 for proof of (6.42).

Note that at the first iteration when the estimate of the inference is not yet available, the SIC cannot be carried out. This also applies to the cancellation scheme presented in Section 6.5.4. One solution to this problem is to apply MMSE filtering without SIC at the first iteration to obtain an initial estimate of interference.

6.7.2   Iterative Trellis Search Equalization

From the model described in (6.30), we may formulate reduced-complexity time-domain equalization schemes based on the optimal maximum *a posteriori* (MAP) estimation implemented using the BCJR algorithm described in Chapter 4 in section 3.2.4. In particular, we will focus on the iterative trellis search equalization (ITSE) for the MIMO channels with delay spread as proposed by de Jong and Willink [33]. The ITSE scheme is based on the M-BCJR algorithm [49]. This algorithm uses the trellis search with a reduced number of states, namely M number of states; the parameter M can be chosen to be much lower than the maximum number of possible states, depending on the number of multipaths N_h, the modulation order M_c, and the number of transmit antennas n_t. Furthermore, ITSE can be combined with multilevel map partitioning of the QAM modulation as described in Section 6.5.3, in which case the complexity of the proposed ITSE algorithm can be made independent of the modulation order.

MAP Equalization   For a positive definite $\mathbf{R}$ matrix in (6.30), we may use the Cholesky decomposition to write $\mathbf{R} = \mathbf{L}^\dagger \mathbf{L}$, where $\mathbf{L}$ is an $N_s \times N_s$ lower-triangular matrix with lower bandwidth $L_b = N_h n_t - 1$. The noise-whitened signal can be obtained using a premultiplication of the signal in (6.30) by $\mathbf{L}^{-H}$, which yields

$$\tilde{\mathbf{y}} = [\tilde{y}_1, \cdots, \tilde{y}_{N_s}]^T = \mathbf{L}^{-H}\mathbf{y} = \mathbf{L}\mathbf{x} + \tilde{\mathbf{v}} \quad (6.43)$$

where the noise vector $\tilde{\mathbf{v}} = [v_1, \cdots, v_{N_s}]^T = \mathbf{L}^{-H}\mathbf{v}$, whose elements are independent complex Gaussian variables with variance N_0. Denoting the $(i, i-j)$th element of $\mathbf{L}$ by $l_{i,j}$. The element $\tilde{y}_i$ can be rewritten as

$$\tilde{y}_i = l_{i,0} x_i + \sum_{j=1}^{L_b} l_{i,j} x_{i-j} + \tilde{v}_i \tag{6.44}$$

The complexity of the optimal BCJR equalization/detection on the trellis representing the wideband MIMO channel process is proportional to the number of possible state transitions, which is $2^{N_h n_t M_c}$. Refer to Chapter 4 for a detailed description of the BCJR algorithm. In this context, we note the following low-complexity schemes:

- The M-BCJR algorithm proposed by Fragouli et al. [48], Franz and Anderson [49], and Vithanage et al. [161] substantially reduces this complexity by processing fewer states ($M < 2^{N_h n_t M_c}$) than the total number of states in the trellis. Here the state transition is associated with the number of transmit antennas n_t and the modulation size M_c; thus, the the number of active states at each stage in the forward recursion of the M-BCJR algorithm is $M \cdot 2^{n_t M_c}$.

- The ITSE scheme proposed by de Jong and Willink in [33], which is a symbol-by-symbol approach, is based on a trellis representation of the channel model in (6.44) that is independent of number of transmit antennas. The number of active states at each stage in the forward recursion of the M-BCJR algorithm of this symbol-by-symbol ITSE scheme is on the order of $M \cdot 2^{M_c}$.

ITSE Proposed in (33) In order to describe the state space of the trellis representing the channel process (6.44), we first define the following terms:

- $\mathbb{X}$ is the state space of the trellis described by a set of the elements which are indexed by the integer $m = 0, \cdots, 2^{L_b M_c} - 1$. The actual state of the process at stage i is represented by $X_i \in \mathbb{X}$.

- $\mathbb{U}$ is the state transition space consisting of all possible transitions between states in $\mathbb{X}$. The elements of $\mathbb{U}$ are indexed by the integer $u = 0, \cdots, 2^{(L_b+1)M_c} - 1$.

Each of the possible paths through the trellis is characterized as follows:

- The path starts in the initial state $X_0 = 0$ and may end in any of the possible $2^{L_b M_c}$ terminal states.

- The L_b symbols in the state vector, given that the process is in state m, are denoted by $\hat{x}_1(m), \cdots, \hat{x}_{L_b}(m)$.

- For each state transition in the trellis, we have starting and ending states $X^S(u)$ and $X^E(u)$, respectively, as well as the input symbol causing the transition, $\hat{x}_0(u)$. The M_c bits associated with $\hat{x}_0(u)$ are represented by $\hat{\mathbf{c}}(u) = [\hat{c}_1(u), \cdots, \hat{c}_{M_c}(u)]^T$.

Then we have the following computations for the M-BCJR algorithm:

1. The extrinsic LLRs of the transmitted bits $c_{i,k}$ are computed as

$$\Lambda(c_{i,k}; e) = \log \frac{\displaystyle\sum_{u:\hat{c}_k(u)=+1} \exp \mu_i(u)}{\displaystyle\sum_{u:\hat{c}_k(u)=-1} \exp \mu_i(u)} - \Lambda(c_{i,k}; a) \qquad (6.45)$$

2. The metric $\mu_i(u)$ represents the log-likelihood of u at stage i, given $\mathbf{y}$ and the *a priori* LLR $\Lambda(:, a)$; it can be written as

$$\mu_i(u) = A_{i-1}(X^S(u)) + B_i(X^E(u)) + C_i(u). \qquad (6.46)$$

where the forward and backward state metrics, $A_i(\cdot)$ and $B_i(\cdot)$, are computed using forward and backward recursions, respectively, and the transition metric $C_i(u)$ is evaluated as

$$C_i(u) = -\frac{1}{N_0} \left| y_i - l_{i,0}\hat{x}_0(u) - \sum_{j=1}^{L_b} l_{i,j}\hat{x}_j(X^S(u)) \right|^2 + \sum_{j:\hat{c}_j(u)=+1} \Lambda(c_{i,j}; a).$$
$$(6.47)$$

In the ITSE, we note the following properties:

- The performance of the M-BCJR algorithm is identical to that of the BCJR algorithm only if $M = 2^{L_b M_c}$.
- In practice, near-optimal performance is obtained when M is only a small fraction of $2^{L_b M_c}$.
- The efficiency of the M-BCJR algorithm depends on the spatiotemporal correlation between the symbols. When the channel is highly correlated, the diagonal elements of $\mathbf{L}$ tend to zero, in which case near-optimal performance can be achieved only by a exhaustive trellis search,(i.e., for very large values of M).
- The complexity per bit of the symbol-by-symbol M-BCJR algorithm can be reduced further by the use of multilevel bit mappings.

Applying multilevel mapping (MLM) for the ITSE with MLM reduces the complexity per bit of the forward recursion almost independently of the constellation size. In particular, exploiting the properties of multilevel bit mappings presented in Section 6.5.3, the trellis structure representation of the channel model (6.44) can be simplified such that the trellis has only 2 bits on each branch. This means that there will be only four branches at each node, regardless of the modulation but $M_c/2$ intermediate stages need to be evaluated [36]. If MLM is not exploited, then there will be M_c bits on each branch and 2^{M_c} branches in each node–; however,

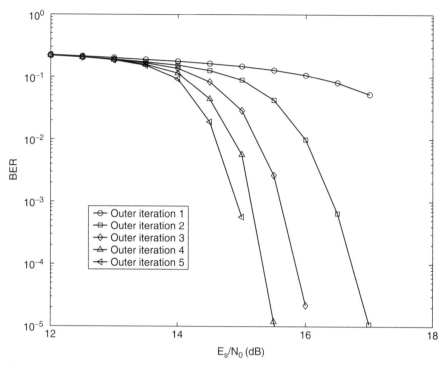

Figure 6.8 Error performance of an 8×8 ST-BICM MLC-ITS system. QAM modulation and list size $M = 32$. The channel code is a turbo code of average rate 1/2 and memory 3.

there are no intermediate stages to be encountered. See [33] for equations to derive the intermediate stages.

6.7.3 Simulation for Wideband Turbo-MIMO

Experiment 1: Performance of MMSE and the ZF MIMO Equalizer with Wideband Channels For the simulations, we consider $n_t = n_r = 4$ and two equal-power mulitpath rays. We use turbo and convolutional codes. The outer codes considered in the simulation are 4-state, rate-1/2 turbo and convolutional codes. QPSK and 16-QAM modulations are considered. Figures 6.9 and 6.10 show BER versus SNR performance of the proposed MIMO equalizers with convolutional and turbo codes, respectively. The noise variance is assumed to be 1. Consequently, the SNR at the jth receive antenna is defined as $\text{SNR}_j = \frac{\rho}{n_t} \sum_{i=1}^{n_t} E|h_{ij}|^2$, where ρ is the total transmitted power. Because of the local stationarity, the SNRs at all receiver antennas are the same. We also assume perfect channel knowledge at the receiver. Figure 6.9 shows the performance of the proposed MIMO equalizer with ZF and MMSE criteria with convolutional codes. The performance of the MMSE equalizer is significantly greater than that of the ZF receiver. Figure 10 shows the

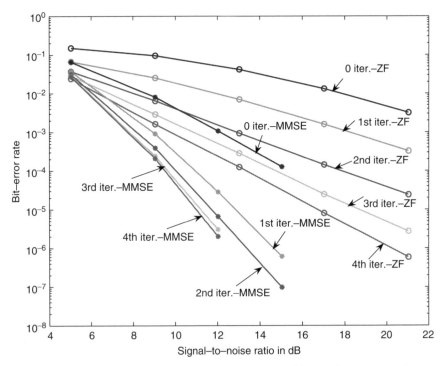

Figure 6.9 BER versus SNR for various iterations with MMSE and ZF equalizers. We use convolutional code with a block size of 400 bits, rate $R = 1/2$, constraint length 3, and QPSK modulation.

performance of the MIMO-MMSE equalizer with turbo codes. A significant gain is obtained with the use of turbo codes.

Experiment 2: Performance of ITSE with Wideband Channels In this section, we compare the performance of narrowband and wideband systems [36] using the ITSE algorithm.

Figure 6.11 shows the simulated error performance of ITSE in a $n_t = n_r = 2$ turbo-MIMO system employing QPSK modulation. The channel model chosen for this simulation is a three-tap multipath channel ($N_h = 3$) with tap power distribution [0.25, 0.50, 0.25]. The ITSE performance is compared to that of the BCJR algorithm and MF and the matched-filter bound (MFB). It can be seen from Figure 3 that ITSE performance is within 3 and 0.5 dB from BCJR performance for M = 4 and M = 32, respectively. Note that M = 4 and M = 32 are only small fractions of the size of the state transition space, which is 4096.

Figure 6.12 shows the simulated error performance of ITSE in a $n_t = n_r = 4$ [Turbo-MIMO system operating in a five-tap channel ($N_h = 5$) with tap power distribution [0.11, 0.22, 0.33, 0.22, 0.11]. In the figure, ITSE performance is compared to the MFB as well as to a variant of ITSE which employs the conventional

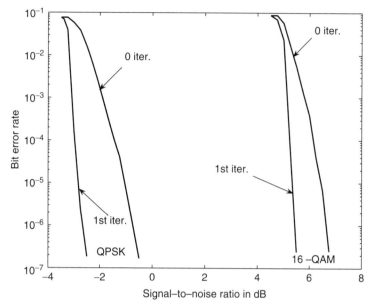

Figure 6.10 BER versus SNR with an MMSE equalizer. We use turbo code with a block size of 16,000 bits, rate $R = 1/2$, and QPSK and 16-QAM modulation.

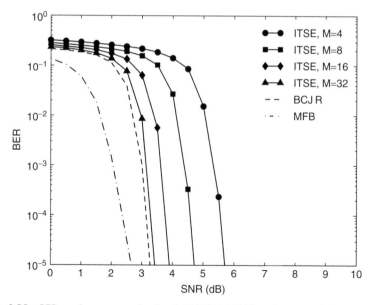

Figure 6.11 BER performance of a 2×2 ST-BICM MIMO system employing the wideband ITS detector for QAM modulations. The channel code is a turbo code of average rate $1/2$ and memory 3.

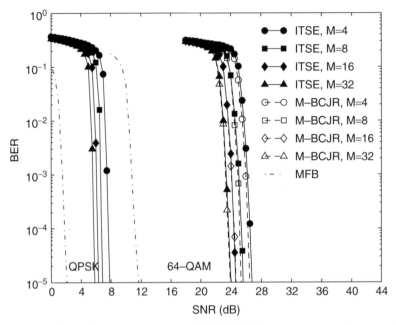

Figure 6.12 BER performance of a 4 × 4 ST-BICM MIMO system employing the wide-band ITSE detector for QAM modulations. The channel code is a turbo code of average rate 1/2 and memory 3.

M-BCJR algorithm [161]. The M-parameter of the ITSE scheme was varied from 4 to 32. For $n_t = n_r = 4$ and 64-QAM with $N_h = 5$, a value of M = 4 corresponds to 10^{-36} times the size of the state transition space. The performance degradation with respect to the MFB is approximately 15 dB for ITSE with M = 4 in the case of 64-QAM modulation. This can be reduced by increasing M, at the cost of increased complexity.

6.8 SUMMARY

In this chapter, a number of reduced-complexity detection schemes for ST-BICM MIMO systems were described. The ST-BICM systems have a structure similar to that of the Turbo-BLAST systems, but distinguish themselves by using a single channel code and interleaver similar to those of BICM at the transmitter and replacing the layered space-time code structures and their corresponding iterative (turbo) processing at the receiver. They generally have considerably better performance than noniterative MIMO systems, although existing schemes do not operate as close to the Shannon limit as single-antenna systems with similar channel codes. In the meantime, their computational complexity is reasonably low. In contrast to the optimal APP detector, the reduced-complexity schemes discussed in this chapter

(LSD, ITSE, and SIC-MMSE detection) do not exhibit the same exponential growth in complexity as the number of transmit antennas increases. The complexity per bit of the multilevel mapping ITSE detector is practically independent of the number of antennas and the signal constellation size. Similar to the LSD scheme, it also offers the possibility of trading off performance for lower complexity by changing the list size parameter of the algorithm. The SIC-MMSE scheme is quite different from the LSD, ITS, and multilevel map-ITS in that it is based on an adaptive filtering approach but has a performance comparable to that of multilevel map-ITS at roughly the same computational cost.

Although the Turbo-MIMO architectures discussed herein are further away from theoretical capacity limits than single-antenna wireless systems with similar channel codes, they can provide considerably higher spectral efficiency.

Finally, we described the extension of the Turbo-MIMO concept in the frequency-selective channel. In particular, some reduced-complexity time-domain equalization schemes, namely, the MMSE-based joint equalization, SIC schemes, and ITSE have been proposed. These schemes can mitigate the adverse effects of ISI and multiple-antenna interference with affordable complexity, which is polynomial in the number of transmit antennas and the channel memory length, and which does not grow with constellation size when they are combined with multilevel map detectors. This is a major improvement over BCJR equalization, as well as over some existing suboptimal SISO MIMO equalization schemes. Furthermore, the ITSE scheme offers the possibility of trading lower complexity for improved performance by changing its M-parameter. This can be utilized to meet specific system requirements.

APPENDIX 6.1

The sphere decoding distance $\gamma(\mathbf{x}) = \|\mathbf{y} - \mathbf{H}\mathbf{x}\|^2 = (\mathbf{x} - \hat{\mathbf{x}})^\dagger \mathbf{H}^\dagger \mathbf{H}(\mathbf{x} - \hat{\mathbf{x}})$ can be expanded as follows:

$$\gamma(\mathbf{x}) = [(x_1 - \hat{x}_1), (x_2 - \hat{x}_2), (x_2 - \hat{x}_3)]$$

$$\begin{bmatrix} l_{1,1} & 0 & 0 \\ l_{1,2} & l_{2,2} & 0 \\ l_{1,3} & l_{2,3} & l_{3,3} \end{bmatrix} \begin{bmatrix} l_{1,1} & l_{1,2} & l_{1,3} \\ 0 & l_{2,2} & l_{2,3} \\ 0 & 0 & l_{3,3} \end{bmatrix} \begin{bmatrix} (x_1 - \hat{x}_1)^* \\ (x_2 - \hat{x}_2)^* \\ (x_2 - \hat{x}_3)^* \end{bmatrix}$$

$$= |l_{11}(x_1 - \hat{x}_1)|^2 \tag{6.48}$$

$$+ \left| l_{22}(x_2 - \hat{x}_2)^2 + l_{12}(x_1 - \hat{x}_1) \right|^2 + \left| l_{33}(x_3 - \hat{x}_3)^2 + \sum_{j=1}^{2} l_{3j}(x_j - \hat{x}_j) \right|^2$$

which can be evaluated in a symbol-by-symbol manner starting with the first symbol, x_1, and proceeding to x_3 as follows:

$$\gamma_1 = |l_{11}(x_1 - \hat{x}_1)|^2$$

$$\gamma_2 = \gamma_1 + \left| l_{22}(x_2 - \hat{x}_2) + \sum_{j=1}^{1} l_{2j}(x_j - \hat{x}_j) \right|^2$$

$$\gamma_3 = \gamma_2 + \left| l_{33}(x_3 - \hat{x}_3) + \sum_{j=1}^{2} l_{3j}(x_j - \hat{x}_j) \right|^2 \qquad (6.49)$$

$$\gamma(\mathbf{x}) = \gamma_3$$

APPENDIX 6.2

Consider the following model:

$$\mathbf{y} = \mathbf{A}_i \mathbf{x}_i + \mathbf{B}_i \mathbf{x}_{int} + \mathbf{u}_i \in \mathbb{C}^{N_s \times 1}, \qquad (6.50)$$

Given (6.50), we wish to find the optimum weight vectors $\mathbf{W}_k$ and u_k by minimizing the cost (convex) function:

$$\mathbf{W}_{i,opt} = \arg \min_{\mathbf{W}_i} E\left\{ |\hat{\mathbf{x}}_i - \mathbf{x}_i|^2 \right\} \qquad (6.51)$$

where the expectation is taken over the statistics of both channel noise and data sequence.

Proof. The cost function is written as

$$
\begin{aligned}
F &= E\left\{ |\mathbf{x}_i - \hat{\mathbf{x}}_i|^2 \right\} \\
&= E\left\{ \mathbf{W}_i [\mathbf{A}_i \mathbf{x}_i + \mathbf{B}_i (\mathbf{x}_{int} - \hat{\mathbf{x}}_{int}) + \mathbf{u}_i][\mathbf{A}_i \mathbf{x}_i + \mathbf{B}_i (\mathbf{x}_{int} - \hat{\mathbf{x}}_{int}) + \mathbf{u}_i]^{\dagger} \mathbf{W}_i^{\dagger} \right\} \\
&\quad - 2\mathbf{W}_i \mathbf{x}_i [\mathbf{A}_i \mathbf{x}_i + \mathbf{B}_i (\mathbf{x}_{int} - \hat{\mathbf{x}}_{int}) + \mathbf{u}_i] + \mathbf{x}_i \mathbf{x}_i^{\dagger}
\end{aligned} \qquad (6.52)
$$

We use standard minimization techniques to solve the optimization problem formulated in (6.50). By setting $\frac{\partial F}{\partial \mathbf{W}_i} = 0$, we get

$$\mathbf{W}_i \left[\mathbf{A}_i E\{\mathbf{x}_i \mathbf{x}_i^{\dagger}\} \mathbf{A}_i^{\dagger} + \mathbf{B}_i E\{(\mathbf{x}_{int} - \hat{\mathbf{x}}_{int})(\mathbf{x}_{int} - \hat{\mathbf{x}}_{int})^{\dagger}\} \mathbf{B}_i^{\dagger} + E\{\mathbf{u}_i \mathbf{u}_i^{\dagger}\} \right] = \mathbf{A}_i \mathbf{x}_i \mathbf{x}_i^{\dagger} \qquad (6.53)$$

$$\mathbf{W}_i \left[\sigma_x^2 \mathbf{A}_i \mathbf{A}_i^{\dagger} + \mathbf{B}_i E\{(\mathbf{x}_{int} - \hat{\mathbf{x}}_{int})(\mathbf{x}_{int} - \hat{\mathbf{x}}_{int})^{\dagger}\} \mathbf{B}_i^{\dagger} + \sigma^2 \mathbf{I} \right] = \sigma_x^2 \mathbf{A}_i$$

where $E\{\mathbf{x}_i \mathbf{x}_i^{\dagger}\} = \sigma_x^2 \mathbf{I}$ and $E\{\mathbf{u}_i \mathbf{u}_i^{\dagger}\} = \sigma^2 \mathbf{I}$. We obtain the following solutions to the weight matrix:

$$\hat{\mathbf{W}}_i = \left(\mathbf{A}_i \mathbf{A}_i^{\dagger} + \frac{\mathbf{B}_i \mathcal{E}\{(\mathbf{x}_{int} - \hat{\mathbf{x}}_{int})(\mathbf{x}_{int} - \hat{\mathbf{x}}_{int})^{\dagger}\} \mathbf{B}_i^{\dagger}}{\sigma_x^2} + \frac{\sigma^2}{\sigma_x^2} \mathbf{I} \right)^{-1} \mathbf{A}_i \qquad (6.54)$$

With an increasing number of iterations, we assume that in the limit, $\hat{\mathbf{x}}_{int}$ tends to the true value, $\mathbf{x}_{int}$. In this case, we may ignore the term $\mathcal{E}\{(\mathbf{x}_{int} - \hat{\mathbf{x}}_{int})(\mathbf{x}_{int} - \hat{\mathbf{x}}_{int})^{\dagger}\}$; then the MMSE soft estimation of the desired signal is given by

$$\hat{\mathbf{x}}_{i,\text{MMSE}} = \left(\sum_{j=1}^{n_r} [\mathbf{H}_{ij}]^{\dagger}[\mathbf{H}_{ij}] + (\sigma^2/\sigma_x^2)\mathbf{I} \right)^{-1} \mathbf{y} \in \mathbb{C}^{N_s \times 1}, i = 1, \ldots, n_t. \quad (6.55)$$

□

BIBLIOGRAPHY

[1] K. Abend and B. D. Fritchman, "Statistical detection for communications channels with inter-symbol interference," *Proceedings of the IEEE*, vol. 58, no. 5, pp. 779–785, May 1970.

[2] T. Akimichi and Y. Sheena, "Distribution of eigenvalues and eigenvectors of Wishart matrix when the population eigenvalues are infinitely dispersed and its application to minimax estimation of covariance matrix," *Journal of Multivariate Analysis*, vol. 94, issue 2, pp. 271–299, June 2005.

[3] S. Alamouti, "A simple transmitter diversity scheme for wireless communication," *IEEE Journal on Selected Areas in Communications*, vol. 16, no. 8, pp. 1451–1458, Oct. 1998.

[4] N. Al-Dhahir, "Overview and comparison of equalization schemes for space-time-coded signal with application to EDGE," *IEEE Transactions on Signal Processing*, vol. 50, no. 10, pp. 2477–2488, Oct. 2002.

[5] N. Al-Dhahir and A. H. Sayed, "The finite-length multi-input multi-output MMMSE-DFE," *IEEE Transactions on Signal Processing*, vol. 48, no. 10, pp. 2921–2936, Oct. 2000.

[6] S. L. Ariyavisitakul, "Turbo space-time processing to improve wireless channel capacity," *IEEE Transactions on Communications*, vol. 48, no. 8, pp. 1347–1359, Aug. 2000.

[7] S. L. Ariyavisitakul, J. H. Winters, and I. Lee, "Optimum space-time processors with dispersive interference: Unified analysis and required filter span," *IEEE Transactions on Communications*, vol. 47, no. 7, pp. 1073–1083, July 1999.

[8] E. Ayanoglu, K. Y. Eng, M. J. Karol, Z. Liu, P. Pancha, M. Veeraraghavan, and C. B. Woodworth, "Mobile infrastructure," *Bell-Labs Technical Journal*, pp. 143–161, Autumn 1996.

[9] L. Bahl, J. Cocke, F. Jelinek, and J. Raviv, "Optimal decoding of linear codes for minimizing symbol error rate," *IEEE Transactions on Information Theory*, vol. 20, issue 2, pp. 284–287, Mar 1974.

Space-Time Layered Information Processing for Wireless Communications,
By Mathini Sellathurai and Simon Haykin
Copyright © 2009 John Wiley & Sons, Inc.

[10] P. Balaban and J. Salz, "Optimum diversity combining and equalization in digital data transmission with applications to cellular mobile radio-part I: Theoretical considerations," *IEEE Transactions on Communications*, vol. 40, no. 5, pp. 885–894, May 1992.

[11] S. Baro, G. Bauch, A. Pavlic, and A. Semmler, "Improving BLAST performance using space-time block codes and turbo decoding," *IEEE Global Telecommunications Conference, Globecom'00*, vol. 2, pp. 1067–1071, Nov. 2000.

[12] G. Bauch and N. Al-Dhahir, "Reduced-complexity space-time turbo-equalization for frequency-selective MIMO channels," *IEEE Transactions in Wireless Communications*, vol. 1, no. 4, pp. 819–828, Oct. 2002.

[13] P. A. Bello, "Characterization of randomly time-variant channels," *IEEE Transactions on Communication Systems*, vol. CS-11, no. 4, pp. 360–393, 1963.

[14] S. Benedetto, D. Divsalar, G. Montorsi, and F. Pollara, "A soft-input soft-output maximum a posteriori (MAP) module to decode parallel and serial concatenated codes," *TDA Progress Report* 42-127, Nov. 1996.

[15] S. Benedetto, D. Divsalar, G. Montorsi, and F. Pollara, "Analysis, design, and iterative decoding of double serially concatenated codes with interleavers," *IEEE Journal Selected Areas in Communications*, vol. 16, no. 2, pp. 231–244, Feb. 1998.

[16] C. Berrou and A. Glavieux, "Near optimum error-correcting coding and decoding: Turbo codes," *IEEE Transactions on Communications*, vol. 44, no. 10, pp. 1261–1271, Oct. 1996.

[17] C. Berrou, A. Glavieux, and P. Titmajshima, "Near Shannon limit error-correction coding and decoding: turbo codes," *International Conference on Communications*, Geneva, Switzerland, pp. 1064–1070, May 1993.

[18] E. Biglieri, G. Caire, and G. Taricco, "Limiting performance of block fading channels with multiple antennas," *IEEE Transactions on Information Theory*, vol. 47, no. 4, pp. 1273–1289, May 2001.

[19] E. Biglieri, J. Proakis, and S. Shamai, "Fading channels: information theoretic and communications aspects," *IEEE Transactions on Information Theory*, vol. 44, no. 6, pp. 2619–2692, Oct. 1998.

[20] E. Biglieri, G. Taricco, and A. Tulino, "Performance of space-time codes for large number of antennas," *IEEE Transactions on Information Theory*, vol. 48, no. 7, pp. 1794–1803, July 2002.

[21] G. Caire, G. Taricco, and E. Biglieri, "Bit-interleaved coded modulation," *IEEE Transactions on Information Theory*, vol. 44, no. 3, pp. 927–946, May 1998.

[22] R. W. Chang and J. C. Hancock, "On receiver structures for channels having memory," *IEEE Transactions on Information Theory*, vol. IT-12, no. 4, pp 463–468, Oct. 1966.

[23] I. Chatzigeorgiou, M. Rodrigues, I. Wassell, and R. Carrasco, "Comparison of convolutional and turbo codes for broadband FWA systems," *IEEE Transactions on Broadcasting*, vol. 53, no. 2, pp. 494–503, June 2007.

[24] D. Chizhik, G. J. Foschini, and R. A. Valenzuela, "Capacities of multielement transmit and receive antennas: correlations and keyholes," *IEE Electronic Letters*, vol. 36, no. 13, pp. 1099–1100, June 2000.

[25] W. Choi, K. Cheong, and J. M. Cioffi, "Iterative soft interference cancellation for multiple antenna system," *Wireless Communication and Network Conference*, vol. 1, pp. 304–309, Sept. 23–28, 2000.

[26] C. Chuah, D. Tse, M. Kahn, and R. Valenzuela, "Capacity scaling in MIMO wireless systems under correlated fading," *IEEE Transactions on Information Theory*, vol. 48, no. 3, pp 637–650, Mar. 2002.

[27] S. T. Chung, A. Lozano, and H. C. Huang, "Approaching eigenmode BLAST channel capacity using V-BLAST with rate and power feedback," *IEEE Vehicular Technology Conference, VTC 2001-Fall*, Atlantic City, NJ, pp. 915–919, Oct. 7–11, 2001.

[28] T. M. Cover, "Some advances in broadcast channels" in *Advances in Communications Systems*. New York: Academic Press, 1975, pp. 229–260.

[29] S. Crozier and P. Guinand, "High-performance low-memory interleaver banks for turbo-codes," *IEEE 54th Vehicular Technology Conference, VTC Fall-2001*, Atlantic City, NJ, pp. 2394–2398, Oct. 7–11, 2001.

[30] E. Csiszar and J. Korner, *Information Theory: Coding Theorems for Discrete Memoryless Systems*. New York: Academic Press, 1981.

[31] M. O. Damen, "Joint coding/decoding in a multiple access system: application to mobile communication," Ph.D. Thesis, ENST, Oct. 1999.

[32] Y. L. C. de Jong and T. J. Willink, "Iterative tree search detection for MIMO wireless systems," *IEEE Vehicular Technology Conference*, vol. 2, pp. 1041–1045, Sept. 2002.

[33] Y. L. C. de Jong and T. J. Willink, "Iterative trellis search detection for asynchronous MIMO systems," *IEEE Vehicular Technology Conference, VTC 2003-Fall*, vol. 1, pp. 503–507, Oct. 6–9, 2003.

[34] Y. L. C. de Jong and T. J. Willink, "A reduced-complexity soft-input soft-output detection scheme for wideband MIMO channels," *15th IEEE International Symposium on Personal, Indoor and Mobile Radio Communications, PIMRC 2004*, vol. 2, pp. 810–814, Sept. 5–8, 2004.

[35] Y. L. C. de Jong and T. J. Willink, "Iterative tree search detection for MIMO wireless systems," *IEEE Transactions on Communications*, vol. 53, no. 6, pp. 930–935, June 2005.

[36] Y. L. C. de Jong and T. J. Willink, "Reduced-complexity time-domain equalization for Turbo-MIMO systems," *IEEE Transactions on Communications*, Oct. 2007.

[37] B. Dong and X. Wang, "Sampling-based soft equalization for frequency-selective MIMO channels," *IEEE Transactions on Communications*, vol. 53, no. 2, pp. 278–288, Feb. 2005.

[38] H. El Gamal and E. Geraniotis, "Iterative multiuser detection for coded CDMA signals in AWGN and fading channel," *IEEE Journal on Selected Areas in Communications*, vol. 18, no. 1, pp. 30–41, Jan. 2000.

[39] H. El Gamal and A. R. Hammons, Jr., "New approach for space-time transmitter receiver design," *Proceedings of the 37th Allerton Conference on Communication, Control, and Computing*, pp. 186–194, Oct. 1999.

[40] H. El Gamal and A. R. Hammons, Jr., "New approach to layered space-time coding and signal processing," *IEEE Transactions on Information Theory*, vol. 47, no. 6, pp. 2321–2334, Sept 2001.

[41] R. B. Ertel, P. Cardieri, K. W. Sowerby, T. S. Rappaport, and J. H. Reed, "Overview of spatial channel models for antenna array communications systems," *IEEE Personal Communications*, pp. 10–27, Feb. 1998.

[42] D. Falconer, S. L. Ariyavisitakul, A. Benyamin-Seeyar, and B. Eidson, "Frequency domain equalization for single-carrier broadband wireless systems," *IEEE Communication Magazine*, vol. 40, pp. 58–66, Apr. 2002.

[43] G. J. Foschini, "Layered space-time architecture for wireless communication in a fading environment when using multi-element antennas," *Bell Labs Technical Journal*, vol. 1, no. 2, pp. 41–59, Autumn 1996.

[44] G. J. Foschini, D. Chizhik, M. J. Gans, C. Papadias, and R. A. Valenzuela, "Analysis and performance of some basic space-time architectures," *IEEE Journal on Selected Areas in Communications*, vol. 21, no. 3, pp. 303–320, Apr 2003.

[45] G. J. Foschini and M. J. Gans, "On limits of wireless communication in a fading environment when using multiple antennas," *Wireless Personal Communications*, vol. 6, pp. 311–335, 1998.

[46] G. J. Foschini, G. D. Golden, R. A. Valenzuela, and P. W. Wolniansky, "Simplified processing for high spectral efficiency wireless communications employing multi-element arrays," *IEEE Journal on Selected Areas in Communications*, vol. 17, no. 11, pp. 1841–1852, Nov. 1999.

[47] G. J. Foschini and R. K. Mueller, "The capacity of linear channels with additive Gaussian noise," *Bell System Technical Journal*, pp. 81–94, Jan. 1970.

[48] C. Fragouli, N. Al-Dhahir, S. N. Diggavi, and W. Turin, "Prefiltered space-time M-BCJR equalizer for frequency-selective channels," *IEEE Transactions on Communications*, vol. 50, no. 5, pp. 742–753, May 2002.

[49] V. Franz and J. B. Anderson, "Concatenated decoding with a reduced-search BCJR algorithm," *IEEE Journal on Selected Areas in Communications*, vol. 16, no. 2, pp. 186–195, Feb 1998.

[50] B. J. Frey and D. J. C. MacKay, "Irregular Turbo codes," *Proceedings of the 37th Allerton Conference on Communication, Control, and Computing*, Illinois, 1999.

[51] R. G. Gallager, *Low-Density Parity-Check Codes*, Cambridge, MA: MIT Press, 1963.

[52] R. G. Gallager, *Information Theory and Reliable Communications*, New York: John Wiley and Sons, 1968.

[53] M. J. Gans et al., "BLAST system capacity measurements at 2.44 GHz in suburban outdoor environment," *IEEE Vehicular Technology Conference*, vol. 1, pp. 288–292, May 2001.

[54] D. Gesbert, H. Bolcskei, D. Gore, and A. Paulraj, "MIMO wireless channels: capacity and performance prediction," *IEEE Global Telecommunications Conference, Globecom'00*, San Francisco, vol. 1, pp. 1083–1087, Nov. 2000.

[55] V. L. Girko, "Circular law," *Theory of Probability and Its Applications*, vol. 29, pp. 694–706, 1984.

[56] G. D. Golden, J. G. Foschini, R. A. Valenzuela, and P. W. Wolniansky, "Detection algorithm and initial laboratory results using V-BLAST space-time communication architecture," *Electronics Letters*, vol. 35, no. 1, pp. 14–15, Jan. 1999.

[57] G. H. Golub and C. F. Van Loan, *Matrix Computations*, 3rd. ed., Baltimore: Johns Hopkins University Press, 1996.

[58] N. R. Goodman, "Statistical analysis based on a certain multivariate complex Gaussian distribution (an introduction)," *Ann. Math. Statistics*, vol. 34, pp. 152–177, 1963.

[59] K. Gracie, S. Crozier, and A. Hunt, "Performance of a low-complexity turbo decoder with a simple stopping criterion implemented on a SHARC processor," *6th International Mobile Satellite Conference, IMSC'99*, Ottawa, Canada, pp. 281–286, June 16–18, 1999.

[60] V. Grenander and J. W. Silverstein, "Spectral analysis of networks with random topologies," *SIAM Journal of Applied Mathematics*, vol. 32, no. 2, pp. 499–519, Mar. 1977.

[61] T. Guess, H. Zhang, and T. Kotchiev, "The outage capacity of BLAST for MIMO channels," *IEEE International Conference on Communications, ICC'03*, pp. 2628–2632, May 11–15, 2003.

[62] J. Hagenauer, "The turbo principle: Tutorial introduction and state of the art," *International Symposium on Turbo Codes*, Best, France, *Sept.* 1997.

[63] L. Hang, C. Jason, and R. S. Cheng, "Low complex turbo space-time code for system with large number of antennas," *IEEE Global Telecommunications Conference, Globecom'00*, San Francisco, vol. 1, pp. 990–994, Nov. 2000.

[64] B. Hassibi, "An efficient square-root algorithm for BLAST," Technical Memorandum, Bell Laboratories, Lucent Technologies, 1999.

[65] B. Hassibi, "High-rate codes that are linear in space and time," Technical Memorandum, Bell Laboratories, Lucent Technologies, 2000.

[66] B. Hassibi and B. Hochwald, "High-rate codes that are linear in space and time," *IEEE Transactions on Information Theory*, vol. 48, no. 7, pp. 1804–1824, July 2002.

[67] S. Haykin, *Adaptive Filter Theory*, 3rd ed., Englewood Cliffs, NJ: Prentice-Hall, 1996.

[68] Simon Haykin, *Communication Systems*, 4th ed., Hoboken, NJ: John Wiley and Sons, 2000.

[69] S. Haykin, "Adaptive digital communication receiver," *IEEE Communications Magazine*, vol. 38, pp. 106–114, Dec. 2000.

[70] S. Haykin, M. Sellathurai, Y. L. C. de Jong, and T. Willink, "Turbo-MIMO for wireless communications," *IEEE Communications Magazine*, vol. 42, pp. 48–53, Oct. 2004.

[71] B. M. Hochwald and T. L. Marzetta, "Unitary space-time modulation for multiple-antenna communications in Rayleigh flat fading," *IEEE Transactions on Information Theory*, vol. 46, no. 2, pp. 543–564, Mar. 2000.

[72] B. M. Hochwald, T. L. Marzetta, and V. Tarokh, "Multiple-antenna channel-hardening and its implications for rate feedback and scheduling," *IEEE Transactions on Information Theory*, vol. 50, no. 9, pp. 1893–1909, Sept. 2004.

[73] B. M. Hochwald and S. Ten Brink, "Achieving near-capacity on a multiple-antenna channel," *IEEE Transactions on Communications*, vol. 51, no. 3, pp. 389–399, Mar. 2003.

[74] W. C. Jakes, Jr., *Microwave Mobile Communications*, New York: John Wiley and Sons, 1974.

[75] J. Jalden and B. Ottersten, "On the complexity of sphere decoding in digital communications," *IEEE Transactions on Signal Processing* vol. 53, no. 4, pp. 1474–1484, Apr 2005.

[76] D.R. Jensen, "Inequalities for joint distributions of quadratic forms," *SIAM Journal on Applied Mathematics*, vol. 42, no. 2, pp. 297–301, Apr 1982.

[77] D. Josson, "Some limit theorems for the eigenvalues of a sample covariance matrix," *Journal of Multivariate Analysis*, vol. 12, pp. 1–18, 1982.

[78] H. Krim and M. Viberg, "Two decades of array signal processing research," *IEEE Signal Processing Magazine*, vol. 13, no. 4, pp. 67–94, July 1996.

[79] P. Lancaster and M. Tismenetsky, *The Theory of Matrices*, 2nd ed., New York: Academic Press, 1985.

[80] V. K. N. Lau, "Channel capacity and error exponents of variable rate adaptive channel coding for Rayleigh fading channels," *IEEE Transactions on Communications*, vol. 47, no. 9, pp. 1345–1356, Sep. 1999.

[81] X. Li, H. Huang, G. J. Foschini, and R. A. Valenzuela, "Effects of iterative detection and decoding on the performance of BLAST," *IEEE Global Telecommunications Conference, Globecom'00*, San Francisco, vol. 2, pp. 1061–1066, Nov. 2000.

[82] Y. Liu, M. P. Fitz, and O. Y. Takeshita, "QPSK space-time turbo codes," *IEEE International Conference on Communications, ICC'00*, pp. 292–296, 2000.

[83] Y. Liu, M. P. Fitz, and O. Y. Takeshita, "Full rate space-time turbo codes," *IEEE Journal on Selected Areas in Communications*, vol. 19, no. 5, pp. 969–980, May 2001.

[84] J. Lodge, P. Hoeher, and J. Hagenauer, "The decoding of multidimensional codes using separable MAP filters," *16th Biennial Symposium on Communications*, Queen's University, Kingston, Canada, pp. 343–346, May 27–29, 1992.

[85] A. Lozano, "Capacity-approaching rate function for layered multiantenna architectures," *IEEE Transactions on Wireless Communications*, vol. 2, no. 4, pp. 616–620, July 2003.

[86] A. Lozano and C. Papadias, "Layered space-time receivers for frequency-selective wireless channels," *IEEE Transactions on Communications*, vol. 50, no. 1, pp. 65–73, Jan. 2002.

[87] D. J. C. MacKay, "Good error-correcting codes based on very sparse matrices," *IEEE Transactions on Information Theory*, vol. 45, no. 2, pp. 399–431, Mar. 1999.

[88] D. J. C. MacKay, S. T. Wilson, and M. C. Davey, "Comparison of constructions of irregular Gallager codes," *IEEE Transactions on Communications*, vol. 47, no. 10, pp. 1449–1454, Oct. 1999.

[89] A. Matache, R. D. Wesel, and J. Shi, "Trellis coding for diagonally layered space-time systems," *International Conference on Communications*, New York, vol. 3, pp. 1388–1392, 2002.

[90] T. Mexia and S. K. Cheng, "Low complexity post-ordered iterative decoding for generalized layered space-time coding systems," *IEEE International Conference on Communications, ICC'01*, Helsinki, Finland, vol. 4, pp. 1137–1141, June 2001.

[91] M. Moher, "Cross-entropy and iterative decoding," Ph.D. Thesis, Carleton University, Ottawa, Canada, May 1997.

[92] M. Moher, "An iterative multiuser decoder for near capacity communications," *IEEE Transactions on Communications*, vol. 46, no. 7, pp. 870–880, July 1998.

[93] W. Mohr and W. Konhauser, "Access network evaluation beyond third-generation mobile communications," *IEEE Communications Magazine*, pp. 122–133, Dec. 2000.

[94] R. A. Monzingo and T. W. Miller, *Introduction to Adaptive Arrays*, New York: John Wiley and Sons, 1980.

[95] A. L. Moustakas, H. U. Baranger, L. Balents, A. M. Sengupta, and S. H. Simon, "Communication through a diffusive medium: coherence and capacity," *Science*, vol. 287, p. 287, Jan. 2000.

[96] A. Narula, M. Trott, and G. Wornell, "Information theoretic analysis of multiple-antenna transmission diversity," *IEEE International Symposium on Information Theory and Its Applications*, Sept. 1996.

[97] A. Narula, M. Trott, and G. Wornell, "Performance limits of coded diversity methods for transmitter antenna arrays," *IEEE Transactions on Information Theory*, vol. 45, no. 7, pp. 2418–2433, Nov. 1999.

[98] S. Ohmori, Y. Yamao, and N. Nakajima, "The future generations of mobile communications based on broadband access technologies," *IEEE Communications Magazine*, vol. 38, pp. 134–142, Dec. 2000.

[99] O. Oyman, U. Nabar, H. Bolcskei, and A. Paulraj, "Tight lower bound on the ergodic capacity of Raleigh fading MIMO channels," *IEEE GLobal Telecommunications Conference, Globecom'02*, Taipei, Taiwan, vol. 2, pp. 1172–1176, 2002.

[100] C. B. Papadias, "On the spectral efficiency of space-time spreading schemes for multiple antenna systems," *33rd Asilomar Conference on Signals, Systems and Computers*, Pacific Grove, CA, pp. 639–643, Oct 24–27, 1999.

[101] C. B. Papadias and G. J. Foschini, "A space-time coding approach for systems employing four transmit antennas," *IEEE International Conference on Acoustics, Speech, and Signal Processing*, vol. 4, pp. 2481–2484, 2001.

[102] C. B. Papadias and G. J. Foschini, "On the capacity of certain space-time coding schemes," *Eurasip Journal on Applied Signal Processing*, vol. 5, pp. 447–458, May 2002.

[103] A. Paulraj, "Diversity techniques," *CRC Handbook on Communications*, ed. J. Gibson, Boca Raton, FL: CRC Press, 11: 213–223, 1996.

[104] A. Paulraj and C. Papadias, "Space-time processing for wireless communications," *IEEE Signal Processing Magazine*, vol. 14, no. 6, pp. 49–83, Nov. 1997.

[105] M. S. Pinsker, *Information and Information Stability of Random Processes*, San Francisco: Holden Day 1964, chapter 10.

[106] J. G. Proakis, *Digital Communications*, 3rd ed., New York: McGraw-Hill, 1995.

[107] T. S. Rappaport, *Wireless Communications: Principles and Practice*, Englewood Cliffs, NJ: Prentice Hall, 1996.

[108] T. Ratnarajah, "Limits of multi-user wireless systems using multiple antennas, scheduling and rate feedback," *39th Asilomar Conference on Signals, Systems, and Computers*, Pacific Grove, CA, Oct. 30–Nov. 2, 2005.

[109] T. Ratnarajah and M. Sellathurai, "Limits of multi-user MIMO systems using cross-layer processing," *International ITG/IEEE Workshop on Smart Antennas*, Duisburg, Germany, Apr. 4–5, 2005.

[110] T. Ratnarajah and M. Sellathurai, "TURBO-MIMO transceiver for frequency-selective wireless channels," *Proceedings of the IEEE Vehicular Technology Conference, VTC 2005-Spring*, Stockholm, vol. 2, pp. 878–881, May 30–Jun. 1, 2005.

[111] T. Ratnarajah and M. Sellathurai, "Iterative layered space-time transceiver for ISI wireless channels," *IEEE International Conference on Acoustics, Speech, and Signal Processing*, Toulouse, France, vol. 4, May 14–19, 2006.

[112] T. Ratnarajah and R. Vaillancourt, "Complex random matrices and Rayleigh channel capacity," *Communications in Information and Systems*, vol. 3, no. 2, pp. 119–138, Oct. 2003.

[113] T. Ratnarajah and R. Vaillancourt, "Complex random matrices and applications," *Mathematical Reports of the Academy of Science of the Royal Society of Canada*, vol. 25, no. 4, pp. 114–120, Dec. 2003.

[114] T. Ratnarajah and R. Vaillancourt, "Jacobians and hypergeometric functions in complex multivariate analysis," *Canadian Applied Mathematics Quarterly*, vol. 12, no. 2, pp. 213–239, Summer 2004.

[115] T. Ratnarajah and R. Vaillancourt, "Complex random matrices and rician channel capacity," *Problems of Information Transmission*, vol. 41, no. 1, pp. 1–22, Jan. 2005.

[116] T. Ratnarajah and R. Vaillancourt, "Eigenvalues and condition numbers of complex random matrices," *SIAM Journal on Matrix Analysis and Applications*, vol. 26, no. 2, pp. 441–456, Jan. 2005.

[117] T. Ratnarajah and R. Vaillancourt, "Complex singular wishart matrices and applications," *Computers and Mathematics with Applications*, vol. 50, no. 3–4, pp. 399–411, Aug. 2005.

[118] T. Ratnarajah and R. Vaillancourt, "Quadratic forms on complex random matrices and multiple-antenna systems," *IEEE Transactions on Information Theory*, vol. 51, no. 8, pp. 2979–2984, Aug. 2005.

[119] G. G. Rayleigh and J. M. Cioffi, "Spatio-temporal coding for wireless channels," *IEEE Transaction on Communications*, vol. 44, no. 3, pp. 357–366, Mar. 1998.

[120] Recommendation ITU-R M.1225, "Guidelines for evaluation of radio transmission technologies for IMT-2000," 1997.

[121] M. C. Reed, P. D. Alexander, J. A. Asenstorfer, and C.B. Schlegel, "Near single user performance using iterative multiuser detection for CDMA with turbo-codes decoders," *International Symposium on Personal, Indoor and Mobile Radio Communications, PIMRC'97*, pp. 750–744, Sept. 1997.

[122] T. Richardson and U. Urbanke, "Efficient encoding of low-density parity-check codes," Technical Memorandum, Bell-Laboratories, Lucent Technologies, 1999.

[123] P. Robertson, and Th. Woerz, "A novel bandwidth efficient coding scheme employing turbo-codes," *IEEE Internaitonal Conference on Communications, ICC*, Dallas, TX, vol. 2, pp. 962–967, June 1996.

[124] W. Roh and A. Paulraj, "MIMO channel capacity for the distributed antenna systems," *IEEE Vehicular Technology Conference, VTC 2002-Fall*, Vancouver, Canada, vol. 2, pp. 706–709, 2002.

[125] G. Ryzhik and A. Jeffrey, *Table of Integrals, Series, and Products*, 5th ed., New York: Academic Press, 1994.

[126] M. Sellathurai, "Turbo-BLAST, a novel technique for multi-transmit and multi-receive wireless communications," PHD Thesis, McMaster University, Canada, Apr. 2001.

[127] M. Sellathurai and G. Foschini, "Stratified diagonal layered space-time architectures: signal processing and information theoretic aspects," *IEEE Transactions on Signal Processing*, vol. 51, no. 11, pp. 2943–2954, Nov. 2003.

[128] M. Sellathurai, P. Guinand, and J. Lodge, "Approaching near-capacity on a multi-antenna channel using successive decoding and interference cancellation

receivers," *International Journal of Communications and Networks*, vol. 5, no. 2, pp. 116–123, June 2003.

[129] M. Sellathurai and S. Haykin, "A nonlinear iterative beamforming technique for wireless communications," *33rd Asilomar Conference on Signals, Systems, and Computers*, Pacific Grove, CA, vol. 2, pp. 957–961, Oct. 1999.

[130] M. Sellathurai and S. Haykin, "Turbo-BLAST: A novel technique for multi-transmit multi-receive wireless communications," *Multiaccess, Mobility, and Teletraffic for Wireless Communications*, vol. 5, pp. 13–24, Dec. 2000.

[131] M. Sellathurai and S. Haykin, "A simplified diagonal BLAST architecture with iterative parallel-interference cancellation," *IEEE International Conference on Communications, ICC'01*, Helsinki, Finland, vol. 10, pp. 3067–3071, June 2001.

[132] M. Sellathurai and S. Haykin, "Turbo-BLAST for wireless communications: theory and experiments," *IEEE Transactions on Signal Processing*, vol. 50, no. 10, pp. 2538–2546, Oct. 2002.

[133] M. Sellathurai and S. Haykin, "Turbo-BLAST: Performance evaluation in correlated Rayleigh-fading environment," *IEEE Journal on Selected Areas in Communications*, vol. 21, no. 3, pp. 340–349, Apr. 2003.

[134] M. Sellathurai and S. Haykin, "T-BLAST for wireless communications: first experimental results," *IEEE Transactions on Vehicular Technology*, vol. 52, no. 3, pp. 530–535, May 2003.

[135] M. Sellathurai and T. Ratnarajah, "Achieving MIMO channel capacity using multirate layered space-time coding architectures," *Information Theory Workshop*, Rotorua, New Zealand, Aug. 2005.

[136] A. M. Sengupta and P. P. Mitra, "Capacity of multivariate channels with multiplicative noise: I. Random matrix techniques and large-N expansions for full transfer matrices," available at http://arxiv.org/abs/physics/0010081.

[137] C. E. Shannon, "A mathematical theory of communication," *Bell System Technical Journal*, vol. 27, pp. 379–423, 623–656, July and Oct. 1948.

[138] D. Shiu, G.J. Foschini, M. J. Gans, and J. M. Kahn, "Fading correlation and its effect on the capacity of multielement antenna system," *IEEE Transactions on Communications*, vol. 48, no. 3, pp. 502–513, Mar. 2000.

[139] S. H. Simon and A. L. Moustakas, "Optimizing MIMO antenna systems with channel covariance feedback," *IEEE Journal on Selected Areas in Communications*," vol. 21, no. 3, pp. 406–417, Apr. 2003.

[140] B. Sklar "Primer on turbo code concept," *IEEE Communication Magazine*, vol. 35, no. 12, pp. 94–102, Dec. 1997.

[141] Special issue of *Proceedings of the IEEE*, vol. 95, July 2007.

[142] Q.H. Spencer et al., "An introduction to the multiuser MIMO downlink," *IEEE Communication Magazine*, pp. 60–67, Oct. 2004.

[143] S. Z. Stambler, "Shannon's theorems for a complete class of discrete channels whose state is known at the output," *Problems of Information Transmission*, no. 11, pp. 263–270, Nov. 1976.

[144] B. Steingrimsson, Z. Luo, and K. M. Wong, "Soft quasi-maximum-likelihood detection for multiple-antenna wireless channels," *IEEE Transactions on Signal Processing*, vol. 51, no. 11, pp. 2710–2719, Nov. 2003.

[145] H. Su and E. Geraniotis, "Space-time turbo codes with full antenna diversity," *IEEE Transactions on Communications*, vol. 49, no. 1, pp. 47–57, Jan. 2001.

[146] V. Tarokh, H. Jafarkhani, and A. R. Calderbank, "Space-time block codes from orthogonal design," *IEEE Transactions on Information Theory*, vol. 45, pp. 1456–1567, July 1999.

[147] V. Tarokh, A. F. Naguib, N. Seshardri, and A. R. Calderbank, "Combined array processing and space-time coding," *IEEE Transactions on Information Theory*, vol. 45, no. 4, pp. 1121–1128, May 1999.

[148] V. Tarokh, N. Seshardri, and A. R. Calderbank, "Space-time codes for high data rate wireless communications: Performance analysis and code construction," *IEEE Transactions on Information Theory*, vol. 44, no. 2, pp. 744–765, Mar. 1998.

[149] E. Teletar, "Capacity of multi-antenna Gaussian channels," *Technical Report., AT&T Bell Laboratories*, Murray Hill, NJ, 1996.

[150] S. Ten Brink and G. Kramer, "Design of repeat-accumulate codes for iterative detection and decoding," *IEEE Transactions on Signal Processing*, vol. 51, no. 11, pp. 2764–2772, Nov. 2003.

[151] D. Tujkovic, "Recursive space-time trellis codes for turbo coded modulation," *IEEE Global Telecommunications Conference, Globecom'00*, San Francisco, vol. 1, pp. 1010–1014, Nov. 2000.

[152] M. C. Valenti, "Iterative detection and decoding for wireless communications," Ph.D. Thesis, Virginia Polytechnic Institute and State University, July 1999.

[153] M. K. Varanasi and T. Guess, "Optimum decision feedback multiuser equalization with successive decoding achieves the total capacity of the Gaussian multiple-access channel," *Asilomar Conference on Signal, Systems and Computers*, Pacific Grove, CA, pp. 1405–1409, Nov. 1997.

[154] S. Verdu and S. Shamai, "Spectral efficiency of CDMA with random spreading," *IEEE Transactions on Information Theory*, vol. 45, no. 2, pp. 622–637, Mar. 1999.

[155] H. Vikalo and B. Hassibi, "On the sphere-decoding algorithm II. Generalizations, second-order statistics, and applications to communications signal processing," *IEEE Transactions on Signal Processing*, vol. 53, no. 8, pp. 2819–2834, Aug. 2005.

[156] H. Vikalo and B. Hassibi, "On the sphere-decoding algorithm I. Expected complexity," *IEEE Transactions on Signal Processing*, vol. 53, no. 8, pp. 2806–2818, 2005.

[157] S. Vishwanath, W. Yu, R. Negi, and A. Goldsmith, "Space-time turbo codes: Decorrelation properties and performance analysis for fading channels," *IEEE Global Telecommunications Conference, Globecom'00*, San Francisco, vol. 1, pp. 1016–1020, Nov. 2000.

[158] R. Visoz and E. Bejjani, "Matched filter bound for multichannel diversity over frequency-selective Rayleigh-fading mobile channels," *IEEE Transactions on Vehicular Technology*, vol. 49, no. 5, pp. 1832–1845, Sept. 2000.

[159] R. Visoz, A. O. Berthet, and S. Chtourou, "A new class of iterative equalizers for space-time BICM over MIMO block fading multipath AWGN channel," *IEEE Transactions on Communications*, vol. 53, no. 12, pp. 2076–2091, Dec. 2005.

[160] A. J. Viterbi, "Error bounds for convolutional codes and an asymptotically optimum decoding algorithm," *IEEE Transactions on Information Theory*, vol. 13, no. 2, pp. 260–269, Apr. 1967.

[161] C. M. Vithanage, C. Andrieu, R. J. Piechocki, and M. S. Yee, "Reduced complexity equalization of MIMO systems with a fixed-lag smoothed M-BCJR algorithm," *6th IEEE International Workshop on Signal Processing Advances in Wireless Communications*, New York, pp. 136–140, June 2005.

[162] X. Wang and V. Poor, "Iterative (TURBO) soft interference cancellation and decoding for coded CDMA," *IEEE Transactions on Communications*, vol. 47, no. 7, pp. 1046–1061, July 1999.

[163] R. D. Wesel, X. Liu, and W. Shi, "Trellis codes for periodic erasures," *IEEE Transactions on Communications*, vol. 48, no. 6, pp. 938–947, June 2000.

[164] J. H. Winters, "Optimum combining in digital mobile radio with cochannel interferences," *IEEE Journal Selected Areas in Communications*, vol. SAC-2, no. 4, pp. 528–539, July 1984.

[165] A. D. Wyner, "Recent results in the Shannon theory," *IEEE Transactions on Information Theory*, vol. IT-20, no. 1, Jan 1974.

[166] M. Zeng, A. Annamalai, and V. K. Bargava, "Harmonization of third-generation mobile systems," *IEEE Communications Magazine*, vol. 38, no. 12, pp. 94–104, Dec. 2000.

[167] H. Zhang and T. Guess, "Asymptotical analysis of the outage capacity of rate-tailored BLAST," *Proceedings of the Global Telecommunications Conference, Globecom'03*, San Francisco, vol. 4, pp. 1797–1801, Dec. 1–5, 2003.

[168] H. Zheng and D. Samardzija, "H.263 video over BLAST wireless test-bed," *35th Annual Conference on Information Sciences and Systems*, Baltimore, Mar. 2001.

Index

Space-Time Layered Information Processing for Wireless Communications,
By Mathini Sellathurai and Simon Haykin
Copyright © 2009 John Wiley & Sons, Inc.